FORENSIC ENGINEERING:

Learning from Failures

Proceedings of a Symposium sponsored by the
ASCE Technical Council on Forensic Engineering
and the Performance of Structures Research Council
of the Technical Council on Research
held in conjunction with the
ASCE Convention in
Seattle, Washington
April 7, 1986

Edited by Kenneth L. Carper

Published by the
American Society of Civil Engineers
345 East 47th Street
New York, New York 10017-2398

Library of Congress Catalog Card No.: 86-70199
ISBN 0-87262-517-6
Manufactured in the United States of America.

FOREWORD

This volume contains the papers submitted for presentation at the Symposium on "Forensic Engineering: Learning from Failures" held at the Spring Convention of the American Society of Civil Engineers in Seattle, Washington on April 7, 1986. The Symposium was organized by the editor at the direction of the two co-sponsors: the ASCE Technical Council on Forensic Engineering (Narbey Khachaturian, Chairman) and the ASCE Performance of Structures Research Council (James Fisher, Chairman). All papers were invited. Final acceptance was based on two reviews, coordinated by the editor. All papers are eligible for discussion in the Journal of Performance of Constructed Facilites and all papers are eligible for ASCE awards.

The Symposium was designed to review the activities and qualifications of the forensic engineer involved in the investigation of failures in constructed facilities. Procedures of failure investigation are discussed along with the importance of data collection and failure information dissemination. The contributions of forensic activities to the civil engineering professions and to society at large through learning from failures are presented. Trends in construction litigation and insurance alternatives are reviewed. Implications for quality assurance/quality control procedures are also discused.

The discipline of Forensic Engineering has achieved considerable visibility and recognition in recent years. In the past, "forensic engineering" has been associated primarily with connotations of litigation activity (i.e., the engineer as "expert witness"). While forensic engineers continue to play an important role in litigation, they are becoming increasingly involved in a wide variety of activities which may have little to do with litigation. Repair and replacement of the deteriorating infrastructure relies often on the skills and knowledge of forensic experts. Adaptive reuse and historic conservation of significant architectural structures depends upon the forensic consultant to assess existing conditions and make recommendations for repair procedures. The forensic engineer is somewhat of a historian, one who must have a working knowledge of construction practices, materials and design standards used in past eras.

Forensic engineers have also recently begun to take a more active role in coordinating the dissemination of information resulting from failure investigation, so that design and construction procedures might be improved. Construction engineering has its foundation in trial-and-error experience. Much of current practice in the construction industry has evolved through the process of "learning from failures." The forensic engineer can contribute significantly to this process, just as the medical pathologist has made significant contributions to advancements in the practice of medicine.

Forensic engineering is an exciting and demanding profession, one which serves society by advancing the art and science of engineering through learning from failures.

Kenneth L. Carper, Editor

CONTENTS

Forensic Engineering

Learning from Failures

WHAT IS A FORENSIC ENGINEER?

By
Joseph S. Ward, P.E., F.ASCE*

Abstract

A forensic engineer is one who is concerned with the relationship and application of engineering facts to legal problems. He is an acknowledged expert who investigates construction-related failures and claims and subsequently determines causation and responsibility. A licensed professional engineer, he frequently will be called upon to mediate disputes, to assist in the resolution of controversies and may appear in court or in arbitration hearings as an expert witness. Above all, he maintains a posture of objectivity in any disputes in the absence of any conflict of interest.

Introduction

Forensic engineering is a specialized engineering discipline, which is not familiar to the lay public. Indeed, many professional engineers are not aware of the specific tasks and essential qualifications of the forensic engineer.

The field of forensic medicine is perhaps better understood as a result of the use of forensic medical experts in televised crime drama. In medicine, the forensic expert performs autopsies on human bodies to determine causation of death. When acting as a detective, the forensic medical expert may assist in the identification of those responsible for that death.

In a similar way, a forensic civil engineer performs "autopsies" on bridges, buildings, dams and other engineered constructed works to find out why they failed. If it is within the defined scope of services, the forensic engineer may render an opinion regarding responsibility for the failure.

The field of forènsic engineering involves many of the same challenges and rewards which are found in the field of forensic medicine.

A comprehensive definition of forensic engineering has been advanced by Milton F. Lunch, Esq., NSPE General Counsel:

*President, Joseph S. Ward, P.E., Inc., Montclair, New Jersey, and past president, American Society of Civil Engineers

> "Forensic engineering is the application of the art and science of engineering in the jurisprudence system, requiring the services of legally qualified professional engineers. Forensic engineering may include investigation of the physical causes of accidents and other sources of claims and litigation, preparation of engineering reports, testimony at hearings and trials in administrative or judicial proceedings, and the rendition of advisory opinions to assist the resolution of disputes affecting life or property."

Forensic engineering is a broad discipline which includes experts working in fields other than structural or civil engineering. Organizations such as the National Academy of Forensic Engineers bring together experts who work in such areas as traffic safety, fire investigation, product liability, etc. The focus of this paper, however, will be on forensic activity related to civil engineering works.

Qualifications of a Forensic Engineer

There are two prime qualities that must be attributed to a forensic engineer. This individual must first of all, be an expert in his field and, secondly, must be impartial during the course of the investigation for causation and responsibility and must "call the shots as they are."

As an expert, the forensic engineer must have a thorough knowledge of the subject under investigation. This knowledge in a specialized area must have been acquired not only by formal education but by many years of practice in this specialty so that the forensic engineer can relate to similar situations in the past for assistance in the determination of causation and responsibility. The forensic engineer must always avoid assignments that differ from his specialized expertise.

Particularly when the forensic engineer is testifying as an expert in legal proceedings, this individual must be licensed as a professional engineer, the only recognized legal status as an engineer. This is an essential qualification when testimony is given as an expert witness in court, arbitration or at public hearings. In addition, the qualifications and credibility of a forensic engineer are enhanced if this individual maintains active membership in recognized engineering societies, particularly those that focus on the specialty areas involved in the investigation. Predominant in this respect is the prior presentation of papers on subjects that are directly related to the specific forensic engineering assignment.

It must be pointed out that there is a difference between a good "designer" and a good "forensic" engineer. When the investigation is approached from the "design process" point of view, the results are not always satisfactory. Therefore, the forensic engineer must be experienced in investigative techniques coupled with a thorough knowledge of design and construction.

As to impartiality in connection with an assignment, the forensic engineer must avoid any conflict of interest that could possibly construe a bias, or advocacy, to one or more parties who were involved in the failure under investigation.

At this point, I would like to define the word "failure." This does not necessarily connotate a catastrophic event. It does not necessarily mean the collapse of a bridge, building, or other constructed facility that has resulted in loss of life or property. It is generally accepted that a failure occurs when a facility does not perform as it was originally intended by the owner, the design professional, or the contractor. It could be a differing site condition that was unexpected by one or more of the project entities; it could be an air conditioning system that does not function properly; or it could be that site drainage is not adequately removing surface runoff. The foregoing are only a few examples that may be considered a failure. It is the function of the forensic engineer to determine why that failure occurred and in some cases, to determine who was responsible for that facility not performing as it was originally intended.

Because failures involve controversy, it is only axiomatic that the work of a forensic engineer is closely associated with litigation. This specialist most often develops the facts associated with a claim, working closely with the attorneys who are representing the plaintiff or the defendants in a legal action. However, it must borne in mind that even though the forensic engineer may be retained by one of the parties in the dispute, this individual must always maintain the professional approach that the forensic engineer is not a client advocate. It must be clearly understood by the client that the results of the investigation may not be favorable to the client's legal position. In fact, a forensic engineer may be extremely helpful to a client if the weak points or uncertainties in the case are pointed out, as well as providing information which may support the position of those who retained him.

Frequently, the causation and/or responsibility for a failure is not always "black and white." There are often many contributing factors to a failure and therefore conflicting opinions are plausible. Each forensic engineer may approach an investigation from a different viewpoint and two or more recognized experts may not always come to the same conclusion. It is for this reason that reports and court testimony may substantiate the positions of opposing parties while each such expert will be convinced that the posture taken is factual and uncontestable. However, the forensic engineer must never be a "hired gun" to bend to the client's wishes.

Investigation

The most important function of the forensic engineer is the conduct of an intensive and thorough investigation. The final conclusions or opinions and the subsequent testimony as an expert witness, are all predicated on the detailed investigation, which must consider every aspect of the failure. The forensic engineer becomes a "Monday morning quarterback" in reviewing all the elements of the project from its conceptual formulation through design and construction. The forensic engineer must thoroughly review all of the documents that were developed throughout the course of the project including not only the contract documents, but the design analysis, construction change orders, engineering reports, correspondence, job memoranda, and daily field reports maintained by the owner's and the contractor's representatives. These also include job progress photographs and particularly photographs that were taken at the time of the failure.

Concerning the latter, if at all possible the forensic engineer should be engaged immediately following the failure event. This individual should be present at the site before any restorative or corrective action is taken so that he may observe, first hand, the materials that may have been the cause of the failure. Two notable examples of this are the collapse of the bridge over the Mianus River in Connecticut and the collapse of the walkway at the Hyatt Regency Hotel in Kansas City. In both of these instances, forensic engineers were on location within hours following these accidents to make detailed examinations and to photograph and sample the debris prior to their removal.

The investigation often requires a "team" of experts, each in a specialized field. For this type of investigation the forensic engineer assumes the role of a coordinator, orchestrating, tabulating and assimilating the work of others and then drawing conclusions from this unified effort.

Report

The results of the forensic engineering investigation are frequently incorporated in a comprehensive engineering report which recites the elements of the investigation, the analysis and evaluation of all of the data that was acquired by the forensic engineer and the conclusions and opinions on the results of the studies. Pertinent back-up material should be appended to the report as a substantiation of the forensic engineer's final conclusions. This report, if favorable to the forensic engineer's client, will probably be distributed to all parties in the litigation and may serve as the basis for the forensic engineer's possible testimony as an expert witness.

There may be occasions when the client does not require the forensic engineer to submit a report and the results of the investigation are orally stated. These results may later be presented in court or arbitration testimony. The client's wishes in this regard should be clearly understood at the outset of the forensic engineer's engagement.

Up to the point of distribution of the forensic engineer's report, this individual is usually serving as a consultant to the attorney representing his client. In this capacity, the forensic engineer may not as yet have been identified as an expert witness in legal proceedings. This individual therefore enjoys the confidentiality privilege of all efforts being considered as attorney work product. As such, a deposition probably will not be taken nor will the forensic engineer's records be subpoenaed by the opposing party in their discovery proceedings.

However, once the forensic engineer's report is in the hands of the opposing parties and this person has been identified as a potential expert witness, he becomes "fair game" to discovery and to being deposed.

The forensic engineer's report can very often be the basis for an out-of-court settlement in which case a trial or arbitration may never ensue. In my experience, I have been involved in literally hundreds of cases over the past 35 years and probably less than 25% have resulted in a trial or arbitration where I have presented expert testimony. Litigation is costly and time consuming. We all know the crowded court calendars that have delayed trials, as well as arbitration proceedings, resulting in litigation that may be conducted many years after the event that precipitated the litigation. Attorneys and their clients usually welcome the opportunity to settle matters as rapidly and judiciously as possible, rather than prolong the legal proceedings by going to court or arbitration.

The Trial

Once the decision is made that the case will be brought to trial or to arbitration hearings, the forensic engineer as an expert witness is highly instrumental in the pleadings of the attorney. The presentation of facts and conclusions by this expert witness is most often the deciding factor in the outcome of the case.

In direct testimony, the expert witness must present his qualifications, factual evidence, and final conclusions in such a manner that he will be "believable" in the eyes of the judge and/or the jury. While the expert's opinions may be contested by experts on the other side, the forensic engineer must present the facts and opinions in a manner that will convince the court that the findings should be accepted, in spite of any contradictory testimony that may be offered by the opposing parties.

During cross-examination, the witness must steadfastly maintain the original technical presentation, even though the opposition will try their best to have their experts cast some possible doubt as to this individual's conclusions. The expert must consistently maintain a professional demeanor and "keep his cool" throughout the course of the cross-examination.

In effect, the expert witness becomes a teacher. In both direct testimony and in cross examination the forensic engineer is addressing the court on all of the elements that led to his conclusions. It may appear to some who have a technical expertise that the explanations may border on a professorial level by detailed explanation of basic engineering concepts. However, the expert witness must present testimony as he would to non-technical laymen, as is usually the case of a judge, a jury or one or more of the arbitrators.

Conclusion

The practice of forensic engineering is an exciting endeavor. As an expert, the forensic engineer utilizes all of his technical resources in the finding of facts. The forensic engineer arrives at final conclusions on the basis of sound engineering fundamentals and from the evidence that has been developed during the investigation. The forensic engineer displays no bias during the course of the assignment. The forensic engineer always conducts himself as a professional in expert witness testimony.

Failure Investigations for
Forensic Engineering

George F. Sowers*

1. Introduction

1.1 Example of a Perverted Investigation

Problems developed during construction of a complex public building: 1) the contractor was months behind schedule, 2) anchor bolts in a main column were several inches out of line, and 3) cracks developed in a stairwell wall at its junction with a floor slab. The contractor contacted the news media, claiming defects in design were delaying construction. To support his claim, he retained a prominent forensic engineer to investigate problems (2) and (3). This forensic engineer made a site inspection and a cursory review of the project plans and issued a statement to the press that called attention to the anchor bolt missalignment and the cracks. He added warning of dire consequences if these problems were caused by certain design deficiencies. However he did not pinpoint any deficiencies yet; he said would look for them if he were retained by the contractor. Meanwhile the contractor requested that the work should be suspended (which would relieve him from the responsibility of being behind schedule).

The owner investigated the problems, finding that the contractor had misaligned the bolts while pouring the column at night. The cracks in the wall were from post tensioning the floor slab and were cosmetic. The contractor failed to obtain a suspend order, and it became obvious that the construction delays were due to his own inadequate effort.

*Senior Vice President, Law Engineering Testing Company, Regents Professor of Civil Engineering, Georgia Institute of Technology, Atlanta, GA

The forensic expert's credibility was damaged and his premature statements to the press had harmed the reputations of the design engineer and the architect.

1.2 Example of an Economical Investigation

A dike for a waste treatment lagoon failed during its initial filling. An erosion hole or pipe developed at the base of the dike, from upstream to downstream, draining the waste into a nearby river. The State Environmental Department was alerted, and expressed grave concern.

The owner, contractor and engineer agreed jointly to investigate the cause, and to take immediate remedial action without attempting to fix blame or financial responsibility until after the repairs were complete. All stages of the investigation were to be witnessed by all parties, the work documented, and the data shared. The cause of failure was found to be an undisclosed layer of peat between borings.

The dike was rebuilt and was operating successfully within 10 days. The regulatory agency was satisfied by the prompt action and assessed no fine.

Eventually the owner, contractor, and engineer admitted that each was partially to blame and they agreed on a reasonable sharing of costs. All benefited except the attorneys.

These examples contrast the two attitudes toward problem investigations. The first investigation started with a biased objective of exploiting the problem. The second began with the common objective of correcting the problem. The first ended with loss of credibility of the forensic investigator and with slurs and potential damage to the reputation of the owner and designers. The second ended with a quick, economical repair and a satisfied owner. The examples illustrate that investigation involves technical, economic, and people problems.

2. Obectives of Investigation

There are typically several objectives for an investigation. They are not always clearly stated and sometimes are not obvious. Some may be self-seeking and even illegitimate (like that of the contractor in the first example) and often they are biased. However, it is necessary to understand them and to define them accurately.

The primary objective of any investigation is to determine the causes of the problem. This begins with a search of the technical causes, but it must include economic, political, social and people causes as well. Unfortunately many investigations are limited to technical issues and the other factors, which often control the technical problems, are ignored.

There are a number of secondary objectives which follow the primary. Most important is to correct the problem. A corollary is to prevent similar problems from developing again. Often a major objective is to establish who should pay and how much, although this is often subverted by substituting ability to pay for a responsibility to pay. Sometimes an objective is to divert attention from more critical issues, as in the opening example. This is rarely a legitimate purpose. Politicians however, frequently employ this tactic.

The relative importance of the different objectives is controlled by two sets of forces. The first are the moral and professional responsibilities of the engineers, architects, contractors, owners and attorneys involved. Some take their responsibilities seriously. Others behave like crocodiles or vultures, preying on misfortune.

The second set of forces are political and economic. Inflammatory news accounts are siezed on by politicians who can benefit from public concern, and who demand accountability and blood. Economic pressures develop when real or imagined losses of money and life are involved and where compensation is either appropriate or demanded by greed.

3. Modes of Investigation

There are two modes for an investigation: 1) cooperative, in which all parties work together and 2) adversarial, with several independent simultaneous investigations, each with different objectives. Often one objective is to fix blame, and another to shift blame to someone else.

The cooperative investigation requires joint effort by the concerned engineers, architects, builders and the owner. Questions of blame and who pays are deferred. The immediate objectives are obtaining the facts and making the analyses to determine cause, corrective measures and procedures for preventing future problems. The advantages are objectivity, speed, and minimium cost. The disadvantage is that one of the parties involved may see that he is a large part of the cause as the investigation progresses. When that happens, he may withdraw from the joint effort and undertake a separate investigation to prepare his defense. Cooperative investigations sometimes lack legal rights, such as to interview reluctant witnesses or review documents held by others, because lawyers for each party are reluctant to give up their privileges for the benefit of all.

The adversarial investigation consists of a number of independent investigations. It is controlled by lawyers and operates under court rules. Fixing blame and determining who pays are often the primary objectives. The technical causes may be important only in reaching these objectives. The purpose of each independent investigation is to gain a advantage for one of the parties involved. Data and observations are usually hidden until a court requires that they be released. Analyses and conclusions reached may never be released unless they are to the sponsor's advantage. Thus some facts and alternative hypotheses remain hidden. The advantage of an adversarial investigation is that legal strategies for obtaining data by court order and for evaluating testimony by cross examination can eventually be employed, but too often at a very late date. The disadvantages of this mode are slow progress, high and duplicated costs, and non-objectivity.

Three factors largely control which mode will be employed: 1) motives of the persons involved, 2) time available and 3) money available for investigation. The motives of investigation are diagnosis, correction and prevention, diversion, blame, compensation and revenge. The first three motives favor a cooperative mode; the remainder, the adversarial mode. Time is important. The more rapidly the investigation moves the less likely that data will be lost or destroyed and the less likely memories will be distorted. Urgent correction requires a quick investigation. The cooperative investigation can be much faster and cheaper than the adversarial. However, expensive failures involving injury or loss of life often favor a slower and more expensive investigation in which all those involved can be encouraged (or intimidated) into contributing to a settlement of claims.

4. General Procedure

Regardless of the mode of the investigation, the proper attitude and methodical procedures will do much to fulfill the legitimate objectives. Impartiality is essential to obtaining correct conclusions. Of course there is always pressure to be partial from who ever is paying the cost of investigation. However, even those who may be responsible for a failure need an impartial evaluation of all the facts. Otherwise if litigation ensues, they cannot prepare a sound defense.

Impartiality, good engineering ethics and professional responsibility require that the investigator begin with no premature conclusions. Particularly, public release of conclusions based on incomplete facts or tentative analyses can permanently damage the reputations of engineers, architects, contractors and owners. Unfortunately those who would divert attention from themselves often use this tactic, as described in the opening example.

Each step of an investigation requires good documentation. Data disappear and memories fade as controversy and legal maneuvering prolong an investigation. Unless the facts are preserved, analyses based on them can be faulty. Moreover, if the problem is litigated, well documented data make better evidence than memories.

The investigation should proceed in an orderly methodical way, like any other scientific or engineering effort. What appears to be irrelevant should be evaluated as well as what is obviously important; the significance of data are not always apparent until all the data are assembled.

5. Study of the Failure

The failure study consists of a number of steps whose sequence depends on the nature of the problem and when the study commences.

5.1 Protecting Evidence

The first step is protecting the physical evidence so it can be properly documented later. Many failures require immediate action to minimize injury, loss of life, and damage to property. These steps necessarily disrupt the site and often destroy valuable information. This cannot be prevented; safety is more important than facts. Instead the investigators must work quickly to gain as much information as possible before it is destroyed. Data are often lost to curious bystanders, news people, and even other investigators. Sometimes those with something to hide intentionally destroy valuable evidence. In so far as possible the site should be secured from all but those who understand the importance of preserving the physical evidence. Physical evidence is damaged by exposure to sunshine, rain, freezing and other elements. To some extent this can be minimized, but seldom prevented entirely. Therefore it is urgent that the physical evidence be studied without delay.

5.2 Obtaining Physical Evidence

As previously mentioned the physical evidence must be documented. Photography using still cameras and video tapes is the first tool. More than one angle and different forms and scales should be used, including aerial, multidirectional, stereo, and close-up as needed. Professional photographers can be employed for overall views. However, photographs made by trained investigators to illustrate their findings provide a more detailed record that is more closely related to other data being obtained. Such personal records appear to have better credibility in court.

Measurements of dimensions of components, angles of slopes elevations and positions of structural components, and sizes of cracks or fissures are essential quantitative data. These are accompanied by field drawn sketches (and also by photographs). Good sketches and photographs complement one another; both are required. Field notes are added that explain the measurements, sketches and photographs. All written materials should include the name of the investigator and the date. Some investigators require the notes be signed and even notarized. This is probably superflous unless a lengthy legal battle is brewing. If samples are made for off-site testing their location and conditions of sampling are documented.

The field work must not endanger any failure victims, nor the investigators. It must minimize further failure and damage to the structure.

5.3 Interviewing Witnesses

Witnesses to the failure and to events leading up to failure provide the second part of the evidence of why a failure occurs. In contrast to the physical evidence, testimony is largely subjective. Observations are interpreted by each person's experience, education and prejudices, and are eroded (or sometimes embellished) by memory.

Informal interviews by two interviewers (for verification) are least likely to intimidate a witnesses. If possible a written statement signed by the witness should summarize the statements. Some attorneys attempt to cross examine witnesses and then obtain notarized statements of their testimony. This is often counter productive. Witnesses often refuse to testify voluntarily in such circumstances, and when compelled, are reluctant to speak freely.

Formal hearings are often not productive and are sometimes counter productive. Shy persons will rarely speak while extroverts babel endlessly, providing little useful information. Politicians use them as a forum for publicity. At one hearing for a dam failure, only the lawyers for those injured spoke. They offered no information, but instead gave impassioned pleas on behalf of their clients. Those with any information kept quiet in front of the hearing inquisitor although they talked later in private with more sympathetic investigators.

5.4 Tests

On site tests usually involve full-size components or materials where transfer to the laboratory is impractical or where engineering behavior might be changed by transport. Such tests are an important tool in many investigations where on-site behavior of a large sample or as built structural system cannot be simulated in the laboratory or by analysis. Of course the part that failed cannot be tested because it has been destroyed. However, testing a similar part built in the same way under the same conditions as that part which failed can provide a insight into inherent weaknesses.

Laboratory tests of on-site samples are useful for small intact components and for testing materials. They are a supplement to field tests, not a substitute.

5.5 Compiling Data

The investigators are often overwhelmed by the number of

observations and the volume of physical data and testimony that are accumulated in the investigation. Computer storage and recall is a practical management tool. Factual data are most readily handled by engineer-devised systems while testimony can be stored in systems developed for litigation.

6. Previous Data

Previous data on the project are an important component of an investigation. Failure often begins with a previous phase of the project: planning, design, construction, maintenance or operation. The evolution of that part of the project which failed is traced from its inception through its operation. Data on project performance during maintenance and operation are compared with the intended or predicted performance from planning and design. Records on project construction and particularly regarding any deviations from design are compiled to evaluate their influence on ultimate performance.

The preceeding events are evaluated. These include environmental factors which could be failure related, such as changes in temperature or fluctuations in ground water level. Of course, the influence of catastrophic events, such as floods and earthquakes are more obvious.

Periodic or continuous measurements of the structure itself may be available, such as pore water pressures in a dam, earthquake-induced acceleration in a building, or flow and pressure in a pipe line. These can sometimes be directly correlated with failure.

6.1 Organization and People Related Events

Organization-related events such as administrative decisions, can have an impact on potential failure. For example, administrative policy discouraged the project designers for Teton Dam from inspecting the site regularly during construction and from taking a first-hand part in construction decision making. In this way foundation preparation decisions were made that weakened the dam's resistance to piping. Failure of an embankment of the Walter Bouldin dam involved fill that was placed between Thanksgiving and New Year's day, a period when weather was marginal for earth filling, when holidays interrupted the momentum of construction and quality control inspection, and when the chief inspector was hospitalized. The failure investigation found other sections of fill placed during this period were incompletely tested. Large volumes of poorly compacted fill placed during this time were found remaining in the embankment.

Because a number of failures have been related to such non-technical events as holidays and to such non-technical administrative policy of restricting design engineer's visit to the project site, investigation of these factors is essential, although difficult.

6.2 Schedule

Eventually a detailed schedule of events preceding failure is developed. This is integrated with the technical data to develop a sequence of factors that could be related to the failure.

7. Analysis

The most demanding part of an investigation is the analysis of data and events. This is an intricate research evaluaton in which may possible causes must be evaluated. Eventually, a hypothesis of failure is developed which must be compatible with the sequence of events, the observed phenonema during failure and the physical data obtained. Much of the physical data (including some weaknesses) will be found to be irrelevant. Other seemingly unimportant factors will be found to be crucial. Several possible hypotheses must be examined. Eventually the more reasonable of these will be proposed as "causes". All concerned must realize that any "cause" is usually a hypothesis that rarely can be fully verified. The most important evidence is usually damaged or destroyed by the failure. Therefore, there is no certainty. It is the investigator's professional responsibility to convey that sense of uncertainty to those who prefer to judge all events in terms of "right" and "wrong" with no room for uncertainty.

8. Organization for Investigation

8.1 Investigative Bodies

Fire marshalls are examples of organizations dedicated to reoccurring failures of one kind. However, engineering failures do not occur with such frequency and are of such diverse nature that a ready-to-go organization rarely exists. Therefore most investigations utilize existing engineering organizations, ad-hoc panels or boards.

Design firms often include a wide range of experienced architects and design and construction engineers that can evaluate the different aspects of a problem or a failure. Moreover they are accustomed to working together as a team. Thus they can undertake a investigation promptly and efficiently. There are disadvantages. The people are design-oriented, thinking in terms of synthesizing a structure system instead of analyzing its performance. Particularly they think in terms of what is required to satisfy a code or to provide a margin of safety against failure rather than what can cause failure. A few designers can think in terms of failure; others can not.

Government agencies are similar to design firms. They often have the wide range of technical expertise necessary. Like the design firms, however, most think in terms of safe design that meets code requirements instead of potential failure. Moreover, most are encumbered by bureaucratic administrative systems that make them slow to respond and inflexible when change is needed.

Consulting firms and a few testing organizations that specialize in problem solving have the analytical philosophy that is valuable in failure investigations. Such firms are usually quick to respond with experienced personnel. Some firms have laboratory facilities and some include personnel from the different disciplines required. Such fims probably come closest of all existing organizations to meeting the needs for an investigation.

Independent panels have been appointed for large, politically sensitive failures such as that of Teton Dam. When the team is appointed for its expertise, as was the case of the Teton panel, and has the back up of technicians and specialists who work under its direction, the panel can be very effective. Unfortunately such panels are usually part-time. Moreover, time is required to assemble such a panel and its team of investigators. Despite these shortcomings such panels have usually been very effective.

8.2 Investigative Team

The selection of individuals for the team largely determines the effectiveness of the investigation.

The principal investigator is the key to the team's success. The principal investigator requires unique qualities. First, he must have experience and knowledge of the technical, economic and human problems which are encountered in failures. Second, he must be adept in coordinating and evaluating the work of different specialists. Finally, he must be experienced in reaching conclusions based on diverse and sometimes conflicting data and opinions.

An administrator is required to handle the multitude of details of directing the effort. He makes all necessary arrangements with the owner, project engineer, and public officials, and handles the financial details. In small investigations this is often a function of the principal investigator. However, he can be easily smothered by such duties.

The investigative staff consists of objective observers,

experienced in making tests, measurements and observations. Some will be engineers or scientists, but most will be technicians.

Consultant specialists advise the principal investigator in directing the work of the investigative staff and make analyses of the information within their specialties. They also work directly with the investigative staff in the field.

A legal advisor should be available to assist the principal investigator in maintaining the rights of all those involved. The advisor helps the investigators secure their information by legal means and to put the information in a form that is legally acceptable in case of litigation. However, it is important that the investigation is not dominated by the needs of future litigation. If it is, objectivity will be lost.

8.3 Tools for Investigation

A number of assets or facilities are needed for a successful investigation. First is the right to question people who have knowledge of what happened and of the circumstances before failure developed. As a last resort, subpeona power may be required, but informal agreements to supply information are usually more productive. A similar right to examine documents is necessary. Free access to project files is best, because it is difficult to define what is relevant and what is not.

The investigation requires an adequate staff, as previously described, as well as the aid of specialists in relevent disciplines. The need for such specialists cannot always be defined in advance. The information that becomes available during the investigation may require special analysis. For example a stain discovered on a concrete wall at Walter Bouldin Dam required analysis by both a biologist and a chemist.

A laboratory with a broad range of testing and analytical

facilities is essential to most investigations. The laboratory staff should be experienced in developing special equipment and test procedures to fit failure study needs.

A legal umbrella of some form is needed to protect the investigative team from undue public pressure and harassment from commentators and reporters. Too often the demand for news interferes with the investigative work. In addition, today's legal environment may require protection of the investigators from lawsuits for damage to reputations as well as to property.

Finally, the investigation requires adequate funding. Unfortunately it is very difficult to estimate the cost in advance. If enough were known about the failure to determine exactly what the investigation should include, the investigation would not be necessary. Never-the-less, a budget is required. However, the investigation sponsor should not be surprised by very large changes.

9. Evaluating and Reporting

Evaluating and reporting requires technical skill as well as the ability to communicate the results. The principal investigator and the consultants provide the team that analyzes the results, proposes hypotheses, and evaluates the whole in order to establish the most reasonable conclusions.

Previous reporting models only fit previous investigations. A new one is required for each project. The written report should use simple language so the widest range of readers, from specialists to the public, can understand as much as possible. The report is illustrated to make understanding easier. Detailed technical and back up information are best reserved for an appendix that can be consulted by those who want more information. The report must be careful to distinguish which are facts and which are conclusions derived from the facts. The reliability of the conclusion should be discussed. The legal advisor reviews the report for language that could have some hidden legal meaning.

Verbal reports are often necessary to satisfy the public and the reporters. These should be in simple, straight forward english and should make clear what is fact, what is speculation and what is a conclusion. If litigation is involved, the topics to be discussed should be reviewed by the legal advisor. However, the investigator, not the lawyer, should make the report.

10. Closing

A failure investigation is one of the most demanding of engineering undertakings. It requires an integrated team headed by a principal investigator with a wide range of skills. The objectives should be defined in advance so that the results will be meaningful. The success of the investigation often depends on gaining information before it is lost, forgotten, or destroyed. The need for objectivity and documentation govern all phases of the work. There are few permanently constituted failure investigative organizations. This is understandable. The irregular failure frequency and the different fields of expertise required form each failure make a highly varied work load that may not be attractive to the highly qualified persons necessary for such work. Moreover, each investigation is unique, requiring a different mix of disciplines. At present the most effective teams are supplied by consulting engineers specializing in trouble shooting or by ad hoc panels assembled for each occurrence.

Regardless, a failure investigation is an exciting challenge for those unafraid of difficult unique experiences.

WHAT TO DO WHEN A FAILURE OCCURS

James M. Fisher, Ph.D., P.E.*

Abstract

When a major building problem involving structural failure occurs, action should begin immediately to determine the damage or failure and detect the underlying problem. This paper will outline the steps and procedures to be followed so that an accurate investigation of the building situation occurs. By following proper procedures, the engineer can help owners, insurance adjusters and legal representatives to determine financial losses accurately and, where appropriate, to recover those losses quickly. The steps outlined in the paper are designed to assist the investigative team to protect evidence and precisely identify the reason for failure.

Introduction

If a catastrophic structural failure occurs and human life is involved, local police or fire department personnel are generally first at the site. They will take charge of the situation. In general, the structural investigative team will not be on site at this time. Certainly the main concern when such a catastrophe occurs is regard for human life. Evidence is often destroyed or damaged to the extent that it is impossible to reassemble the collapsed structure. Fortunately, only a few failures have occurred in the past where human life has been involved.

After the initial efforts are undertaken to rescue human occupants, the control of the failure investigation shifts. Generally, the owner takes charge. If the structure is under construction, the general contractor may take charge.

The owner or his representative will most likely select the principal investigator. Often the initial thoughts of the owner as to the cause of the failure will influence the type of investigator selected. Regardless of whom is selected, the investigator should remember that:

> The primary objective of the investigator is to survey and document the conditions at the site.

This may include the procurement of samples of failed components for subsequent testing and examination. The principal investigator should

* Vice President, Computerized Structural Design, Inc., Milwaukee, WI

constantly remind himself and his team members of their primary objective and not get "hung up" on examining the details which relate to the principal investigator's unique expertise. In this regard, investigators should act more like physicists, rather than engineers; they should look for the raw data rather than items of most interest to the investigator.

The Investigation

A complete investigation should include:

1. The definition of the problem to be undertaken by the investigating team.

2. Acquisition of field and test data.

3. Generation of a failure theory.

4. Analysis and conclusions relating to the failure.

Procedures

The following steps outline the procedures for accurate investigation of building problems involving the structure failures.

As soon as damage or a failure episode is noted, the **owner** should:

1. **Notify** a competent structural engineer experienced in building failure investigations. He should **provide** as much detailed information as possible to the engineer.

2. **Avoid** moving materials or other evidence.

3. **Rope** off the damaged area to prevent entry by unauthorized persons.

4. **Locate and interview** eye witnesses for detailed information.

5. **Order** aerial photographs in the event of a major collapse.

The investigation will be completed by the **structural engineer** who will:

1. **Assemble** an investigation team whose members are experienced in the specific nature of the problem.

2. **Perform** visual and photographic observations on-site before any debris is removed (except where evacuation of personnel is paramount).

3. **Establish** a coordinate scheme for detailing the location of various debris.

4. **Develop** failure hypotheses based on the original structure and conditions.

5. **Test** each failure hypothesis by critical observation. **Prepare** "failure element" sketches and **compile** a detailed photographic log. **Perform** field testing as required.

6. **Remove** samples under prescribed conditions and carefully **identify** exhibits for laboratory testing.

7. **Conduct** eyewitness interviews again.

8. **Undertake** a systematic document retrieval and categorization procedure. **Review** original structural design and **conduct** independent structural analysis.

9. **Examine** all data and **develop** final conclusions.

10. **Prepare** a written report and **present** all pertinent findings.

Summary

It is hoped that the information presented herein will serve as a guide for the first time investigator and serve as a refresher to the experienced investigator.

ARBITRATION: A RISKY METHOD FOR RESOLVING DISPUTES

Edward A. Hannan*

Abstract

Arbitration has been widely praised as a speedy, "inexpensive" alternative to litigation for resolving a wide range of disputes, including disputes arising from construction failures. However, the quid-pro-quo for reducing litigation costs is a waiver of substantial safeguards available to litigants in civil actions. The waiver of procedural safeguards must be taken into account when assessing the "costs" of arbitration. The true "cost" of arbitration may be imposition of an unjust award which cannot be remedied thereafter. Further the "direct costs" of an arbitration proceeding may exceed costs of litigation.

INTRODUCTION

Many construction contract documents routinely contain clauses requiring that disputes be submitted to arbitration for resolution. See, e.g.: AIA Document A-201, AIA Document B141. Arbitration is a method of dispute resolution in which the parties contract to submit a controversy to one or more arbitrators for a final and binding resolution. While arbitration may be appropriate for resolving minor claims not involving substantial dollars, it is questionable whether arbitration is a sound alternative to litigation for resolving claims arising from building failures.

Some contend that the greatest benefit of arbitration is that it provides an alternative method of resolving disputes without the time and expense of litigation. However, time and expense are but two of the considerations that bear on whether arbitration is preferable to litigation in failure disputes. There are significant procedural limitations to the arbitration remedy which pose many significant risks which may exceed its benefits, including: (1) the

*Partner, Godfrey, Trump & Hayes, 250 East Wisconsin Avenue, Suite 1200, Milwaukee, Wisconsin 53202.

inability to join third parties; (2) no right to discover the opponent's position; (3) the absence of formal rules of evidence; (4) the lack of explanation of the reasons for the arbitrators' decision; and (5) significant limitations on the right of judicial review of the award. In short, an agreement to resolve disputes by way of arbitration typically entails a waiver of several significant procedural safeguards which would apply in civil actions.

I. The Risk Of The Inability To Join Third-Parties.

In civil litigation the rules of procedure typically grant the parties a broad right to implead other parties whose conduct may have contributed to a loss. See, e.g.: Rule 14, Federal Rules of Civil Procedure (hereafter cited as "F.R.C.P."). In the context of a claim arising from a building failure, there are typically many potentially culpable parties: contractors, subcontractors, product manufacturers, product suppliers, construction managers, owners and design professionals. Since potential damages in building failure cases are typically high and usually beyond the financial capability of any one defendant, the right to join other parties and their insurers becomes significant.

Because arbitration is a remedy created by contract, the general rule is that parties who have not agreed to submit disputes to arbitration cannot be compelled to do so. See: United Steel Workers v. Warrior & Gulf Navigation Co., 363 U.S. 574, 582 (1960); Alabama Education Assn. v. Alabama Professional Staff Organization, 655 F.2d 607, 609 (5th Cir., 1981). Unless all parties to the design and construction process have mutually agreed to submit all disputes to arbitration, there is a significant risk that the truly culpable party may not be required to join in the proceeding. If persons who are partially or fully responsible for the damages arising from a construction failure are not parties to the proceedings, then their obligations will not be adjudicated or determined in the proceeding itself. Parties who are determined to be at fault in the arbitration proceeding may be able to later sue the other parties; however, such a suit will involve an entire relitigation of the issues previously litigated in the arbitration proceeding. Further, the party bound by the arbitration award may not prevail in a later action due to lapse of time, loss of evidence or intervening bankruptcy of the culpable party who could not be compelled to join the arbitration proceeding. Thus, the inability to join all potentially culpable parties in an arbitration proceeding exposes those who have agreed to arbitration to multiple litigation costs, the risk of inconsistent decisions or awards, and the risk that the truly culpable parties will not bear their fair responsibility for the risk of loss.

II. The Risk Of The Lack Of Discovery In Arbitration.

Discovery is a method of compelling an adverse party to disclose facts within its own knowledge, information or belief or to disclose and produce documents within its possession to enable a party to defend an action. The Federal Rules of Civil Procedure provide that parties may obtain discovery regarding any matter, not privileged, which is relevant to the subject matter involved in the pending action. In submitting a claim to arbitration, the parties may lose the litigation right to fully discover the basis of the opponent's claim.

The pretrial discovery devices available under the Federal Rules of Civil Procedure confer upon litigants a very broad right of access to materials and information necessary to facilitate a fair trial and resolution of disputes upon their merits. The federal pretrial discovery devices are intended to ". . . . make a trial less a game of blind man's bluff and more a fair contest with the basic issues and facts disclosed to the fullest practicable extent." _U.S. v. Proctor & Gamble Co._, 356 U.S. 677, 682-683 (1958). The purposes of pretrial discovery devices in civil actions include:

1. _The Preservation of Information for Trial_. Information necessary to full and fair litigation may disappear, either due to causes wholly legitimate or for reasons suspect, unethical or illegal.

2. _The Identification and Clarification of the Basic Issues Between the Parties_.

3. _The Ascertainment of Facts Relative to the Issues, or the Determination of the Availability of the Existence of Evidence_.

4. _The Avoidance of Protracted Trials Through the Elimination of Uncontroverted Facts_.

5. _The Elimination of Surprise_.

6. The Reduction of Burdens Upon the Courts.

7. The Obtaining of Tactical Advantages. E.g.: the ability to "pin witnesses down" by taking depositions under oath.

In arbitration, parties not only lack an absolute right to discovery, but also discovery is typically limited to depositions. The Uniform Arbitration Act (U.A.A.), adapted in some form by 26 jurisdictions, provides, in part:

> "On application of a party and for use in evidence, the arbitrators may permit a deposition to be taken, in the manner and upon terms designated by the arbitrators, of a witness who cannot be subpoenaed or is unable to attend the hearing." [U.A.A. §7(b)]

Although an arbitrator has the power and discretion to allow the taking of depositions and the inspection of documents, Foremost Yarn Mills, Inc. v. Rose Mills, Inc., 25 F.R.D. 4 (D.C.Pa., 1960) (citing 9 U.S.S §7); Mc Cree v. Superior Court of Los Angeles County, 221 C.A.2d 166, 38 Cal Rptr 346 (1963), parties to arbitration governed by the U.A.A. face the risk that the arbitrator will deny an application for discovery. If the application is denied, there will be a significant loss of access to information needed to eliminate the element of surprise at the hearing.

By way of contrast, the Federal Rules of Civil Procedure provide that "after commencement of the action, any party may take the testimony of any person, including a party, by deposition upon oral examination." F.R.C.P. Rule 30. Under the Federal Rules of Civil Procedure, parties typically do not have to apply to the court for permission to take depositions. Further, in civil actions, parties may discover the opposing party's position not only through depositions, but also through interrogatories (written questions to be answered under oath) (Rule 33, F.R.C.P.), demands for production of documents, tangible things, and entry upon premises for inspection and other purposes (Rule 34, F.R.C.P.), and requests to admit or deny facts (Rule 36, F.R.C.P.). In short, an agreement to resolve a dispute by arbitration entails a significant loss of the right of access to information available in civil proceedings: an arbitration hearing can indeed become "a game of blind man's bluff" rather than ". . . a fair contest with the basic issues and facts disclosed to the fullest practicable extent" prior to the hearing.

III. The Risk Of The Absence Of Formal Rules Of Evidence In Arbitration.

One of the major differences between arbitration and litigation involves the use and application of the rules of evidence. The fundamental purpose of an evidentiary code is to ensure that reliable, trustworthy facts are presented to the finder of facts. Civil actions are always governed by an evidentiary code which defines not only what evidence may be received, but also, the manner in which such evidence is to be introduced. Arbitration proceedings, however, are typically not governed by rules of evidence.

The American Arbitration rules provide that the arbitrator shall be the judge of the relevancy and materiality of offered evidence, and that a conformity to legal rules of evidence shall not be necessary. Atlas Floor Covering v. Crescent Homes and Gardens, Inc., 333 R.2d 194, 166 C.A.2d 211 (1959)(A.A.A. Rule 31). If the arbitration panel lacks an attorney or if the attorney arbitrator has inadequate experience in applying an evidentiary code, then there is a significant risk that irrelevant, inflammatory or otherwise untrustworthy evidence (such as hearsay) may be admitted. Although some courts have held that the refusal to receive evidence is not a ground for relief, Atlas Floor Covering v. Crescent Homes and Gardens, Inc., 333 P.2d 194, 166 C.A.2d 211 (Cal.App., 1959); Peninsula National Banks v. Joseph . Turecamo, 56 N.Y.2d 794, 452 N.Y.S.2d 398, 437 N.E.2d 1155 (N.Y., 1982), the majority rule provides that an award may be vacated if the arbitrator excludes relevant material evidence. Hence, arbitrators tend to admit any and all evidence regardless of its trustworthiness or probative value to ensure that a "full and fair" hearing is not denied. The effect of this practice is that otherwise inadmissible evidence may become the basis for the arbitration award. Considering the significant limitations on the right of judicial review of an arbitration award (See Part V, infra.), parties to an arbitration proceeding expose themselves to a significant risk that their rights will be determined upon inaccurate and untrustworthy evidence.

IV. The Risk Of Lack Of Explanation Of The Basis For The Arbitration Award.

Upon conclusion of the evidence presented by both sides, the arbitration hearing is declared closed and under AAA rules, the arbitrator has 30 days in which to make an award. The award is the final decision determining the submitted dispute.

The American Arbitration Association merely requires that the award be in writing and signed by the arbitrator or a majority of the arbitrators. (Rule 42) The AAA also states that the arbitrator may grant any remedy or relief "which the arbitrator deems just and equitable and within the scope of the agreement of the parties." (Rule 43)

The general rule is that the arbitrators are not required to give reasons for making the awards. Reichman v. Creative Real Estate Consultants, Inc., 476 F.Supp. 1276 (S.D.N.Y., 1979). In Kurt Oban Co. v. Angeles Metal Systems, 573 F.2d 739 (2nd Cir., 1978), the court stated that arbitrators do not have to disclose the basis for their awards, and without any showing that the award was made "in manifest disregard of the law," the courts will not look beyond the terms of the award. If the reasoning can be inferred from the facts, the award should be confirmed. Accord: Reynolds Security, Inc. v. Magguson, 459 F.Wupp. 943 (D.C. Penn., 1978); Maidman v. O'Brien, 473 F.Supp. 25 (S.D.N.Y., 1979); Ulene v. Murray Millman of Cal. Inc., 346 P.2d 494, 175 C.A.2d 655 (1959). Further, courts generally will not attempt to analyze "lump sum" awards to determine if such awards were fair. Ballantine Books, Inc. v. Capitol Distributing Co., 307 F.2d 17, 21-22 (2d Cir., 1962).

In addition to not being required to state the reasons for the award, the highest court of Maine held that arbitrators may grant relief even though it could not or would not have been granted by a court of law or equity. The only requirement is that the award be rational. Cape Elizabeth School Board v. Cape Elizabeth Teacher's Assn., 438 A.2d 239 (Me., 1983).

A California appellate court held that although the arbitrator is not required to state the rationale for the award, it is presumed that all issues submitted to arbitrators have been decided. Lauria v. Soriano, 4 Cal. Rptr. 328, 180 C.A.2d 163 (1960). In California, the party opposing an award on the grounds that an issue was not decided would have to produce evidence to that effect. Since no rationale is required from the arbitrator as to the basis for the award, where is that evidence going to come from?

The Federal Court for the Eastern District of Missouri decided, possibly in light of a holding such as found in Lauria, that the findings of the arbitrator should be clear enough so that a reviewing court can understand the basis for the award. Vulcan-Hart Corp. v. Stove Furnace & Allied Appliance Workers, 576 F.Supp. 394 (E.D. Mo., 1981), aff'd 671 F.2d 1182 (8th Cir.). The Vulcan-Hart decision is clearly a minority decision which places a higher duty on the arbitrator than that imposed by the AAA and most jurisdictions.

As will be more fully explained in the next section, the lack of an explanation of the basis for the arbitrators' award significantly hampers a party who wishes to challenge an award in the courts. The practical effect of a rule which allows arbitrators to decline to explain the reasons for their decisions, is that the decision is generally "final". If that decision rests upon otherwise inadmissable or untrustworthy evidence, then the losing party may be forced to pay an unfair debt and the winning party may obtain an unfair windfall. Where damages are high, as in cases involving construction failures, this risk can well out weigh any benefit derived from a "speedy" arbitration hearing.

V. The Risk Of A Limited Scope Of Judicial Review Of Arbitration Decisions And Awards.

As a general rule, arbitration awards may only be vacated on very narrow grounds. The modern tendency in the courts has been to insulate an arbitration award from judicial scrutiny, provided that the arbitrators appear to be impartial and the hearing itself appears to have afforded a fair opportunity to present evidence. For instance, in Wisconsin, arbitration awards are deemed to be "presumptively valid" and can be disturbed only if the challenging party demonstrates invalidity by clear, satisfactory and convincing evidence. Stradinger v. City of Whitewater, 89 Wis. 2d 19, 37, 277 N.W.2d 827 (1979). Indeed, in Milwaukee Bd. of Schl. Dir. v. Milwaukee Teachers' Educ. Assoc., 93 Wis. 2d 415, 287 N.W.2d 131 (1980), the Wisconsin Supreme Court said the following respecting judicial review of arbitration awards:

> "This court has also stated that it has a 'hands off' attitude towards arbitrators' decisions. . . . This court has said that:
>
> Judicial review of arbitration awards if very limited. The strong policy favoring arbitration as a method

> for settling disputes under collective bargaining agreements requires a reluctance on the part of the courts to interfere with an arbitrator's award upon issues properly submitted. . . .
>
> Thus, the function of the court upon review of an arbitration award is a supervisory one, the goal being merely to ensure that the parties receive the arbitration that they bargained for. . . ." Milw. Pro. Firefighters Local 215 v. Milwaukee, 78 Wis. 2d 1, 21, 22, 253 N.W.2d 481 (1977)." Id., at page 422.

The view of the Wisconsin Supreme Court is consistent with the views of most jurisdictions respecting judicial review of arbitration awards.

The Uniform Arbitration Act, §11, provides that "upon application of a party, the court shall confirm an award, unless within the time limits hereinafter imposed grounds are urged for vacating or modifying or correcting the award, in which case the court shall proceed as provided in Section 12 and 13."

§12 of the U.A.A. states that an award shall be vacated only if:

> "1. the award was procured by corruption, fraud or other undue means;
>
> 2. there was evident partiality by an arbitrator appointed as a neutral or corruption in any of the arbitrators or misconduct prejudicing the rights of any party;
>
> 3. the arbitrators exceeded their powers;
>
> 4. the arbitrators refused to postpone the hearing upon sufficient cause being shown thereof or refused to hear evidence material to the controversy or otherwise so conducted the hearing, contrary to the provision of section 5, as to prejudice substantially the rights of a party; or
>
> 5. there was no arbitration agreement and the issue was not adversely determined in proceedings

> under section 2 and the party did not participate in the arbitration hearing without raising the objection."

§13 of the U.A.A. allows for modifications or correction of the award <u>only</u> where:

> "1. there was an evident miscalculation of figures or an evident mistake in the description of any person, thing or property referred to in the award;
>
> 2. the arbitrators have awarded upon a matter not submitted to them and the award may be corrected without affecting the merits of the decision upon the issues submitted; or
>
> 3. the award is imperfect in a matter of form, not affecting the merits of the controversy."

These grounds for vacation and modification of an arbitration award are the grounds in practically every jurisdiction, even those not adopting the U.A.A. The practical effect of such statutory grounds for attacking the arbitration award is that once the award is made, it is final for all purposes and will be given the force of judgment unless the requirements of §§12 or 13 are met. <u>Such grounds do not go to the merits of the controversy; rather, they go to the arbitrator's conduct</u>. For instance, the Wisconsin Supreme Court has said:

> "The decision of an arbitrator will not be interfered with for mere errors of judgment as to law or fact, but the court will overturn an arbitrator's award if there is perverse misconstruction or positive misconduct plainly established, or if there is a manifest disregard of the law, or if the award itself is illegal or violates strong public policy." <u>Milw. Bd. of Schl. Dirs.</u>, supra, at 422.

It is generally been held that arbitration awards are to be deemed presumptively valid. <u>Parking Unlimited, Inc. v. Monsour Medical Foundation</u>, 445 A.2d 758, 229 Pa. Super. 289 (1982); <u>Wilcox Co. v. Bouramas</u>, 19 Ill. Dec. 641, 392 N.E.2d 198, 73 Ill. App. 3d 1046 (1979). Further, as a general rule, a party who challenges an arbitration award carries a very high burden of proof and, accordingly, a correspondingly high risk that the award will be affirmed. As a general rule, the party challenging the award must prove it is invalid through clear, satisfactory and convincing evidence. See: Seville Intern. Inc. v. Galanti Group, Inc., 63 Ill. Dec. 578, 438 N.E.2d 509, 102 Ill. App. 3d 799 (1982); American Inusco Realty Inc. v. Century 21, 420 N.E.2d 692 (1981); Jones v. Kristad, 97 Cal. Rptr. 100, 19 C.A.3d 836 (1971).

Where the parties failed to have the proceedings transcribed, the risk of an affirmance is particularly high because the arbitrators need not and usually do not state why they made their award and, without a transcript, a reviewing court usually cannot determine whether the challenging party met the burden of proof on appeal. Further, absent a transcript, it is very difficult, if not impossible to clearly prove arbitrator impartiality or misconduct. Indeed, in Texas, the rule is that absent a transcript of the proceedings, a reviewing court must presume that the evidence was adequate to support the award. House Grain Co. v. Obst, 159 S.W.2d 903 (Tex. App., 1983). Ironically, avoidance of the cost of a transcript is sometimes touted as a cost-saving benefit favoring arbitration over a civil trial!

As a general rule, arbitration awards will only be vacated for reasons set forth in the governing arbitration statute. Biller v. David, 37 A.D.2d 954, 326 N.Y.S.2d 220 (19__); Hirt v. Hervey, 578 P.2d 524, 118 Ariz. 543 (App., 1978); Daisy v. Lindy's Coffee Shop, Inc., 397 F.Supp. 767 (D.C. Cir., 1975); Rior v. Allstate Ins. Co., 137 Cal. Rptr. 411, 68 C.A.3d 311 (1977). Parties generally cannot expand the limited jurisdiction of the trial courts to review awards since judicial review is limited to those areas expressed in the statute. Konicki v. Oak Brook Racquet Club, Inc., 65 Ill. Dec. 819, 441 N.E.2d 1333, 110 Ill. App. 3d 217 (1982). Authority is clear that mere errors as to law or fact will not be grounds for vacating an award. Giant Markets, Inc. v. Sigmar Marketing Systems, Inc., 459 A.2d 765, 313 Pa. Super. 115 (1983); W.M. Girvan Inc. v. International Brotherhood of Teamsters, 55 A.D. 2d 746, 389 N.Y.S.2d 445 (1976); Lauria v. Soriano, 4 Cal. Rptr. 328, 180 C.A.2d 163 (1960). Further, courts will typically turn a deaf ear to complaints that the arbitrator's decision rests upon otherwise insufficient or inadmissible evidence, Shammas v. National Telefilm Associates, Inc., 90 Cal. Rptr. 119, 11 C.A.3d 1050 (1970), or that the arbitrator made a "mistake" in calculating the damages. In Marion Mfg. Co. v. Long, 588 F.2d 538 (6th Cir., 1978), the Sixth Circuit Court of Appeals noted that federal courts do not sit "to instruct the arbitrator in the computation of damages" and "even if the result reached by the arbitration panel was not 'equitable' it must be upheld by the courts unless clearly erroneous." Id., at page 541, fn. 3.

The practical effect of the rules strictly limiting the scope of judicial review of arbitration awards and the rules imposing a near insuperable burden of proof upon the challenging party, is to render the arbitrator's decision final in most cases. In cases such as building failure cases, where the damages are high, the risks of loss can be unfairly shifted to nonculpable parties without further redress.

VI. The Direct Costs Of A Complex Arbitration Dispute Can Be Substantial.

The A.A.A.'s administrative fee schedule is as follows:

Amount of Claim	Fee
\$1 to \$20,000	3% (minimum \$200)
\$20,000 to \$40,000	\$600 plus 2% of excess over \$20,000
\$40,000 to \$80,000	\$1000 plus 1% of excess over \$40,000
\$80,000 to \$160,000	\$1400 plus ½% of excess over \$80,000
\$160,000 to \$5,000,000	\$1800 plus ¼% of excess over \$160,000

Under the AAA "administrative fee" schedule, the parties must pay the AAA \$1,500.00 for processing a \$100,000.00 claim. In contrast, a claimant in Wisconsin pays \$45.00 as a filing fee to initiate a civil action for more than \$1,000.00 in Wisconsin's Circuit Courts and it costs \$60.00 to file any federal court action.

In addition to the "administrative fee", parties to an arbitration proceeding typically are required to pay each arbitrator a per diem fee on the order of \$400 per day for proceedings after the second day. In complex arbitration proceedings lasting several weeks or months a three-arbitrator panel, charging $1,200.00 per day, can imposes a staggering cost. On the other hand, the judicial system does not impose a per diem fee for a judge's time.

Next, parties to an arbitration must pay for the rental of the hotel or conference room where the proceeding is held. Of course, governments do not charge a rental fee for the use of a court room.

Finally, whether or not parties resolve their disputes thorugh arbitration or litigation, the direct cost of transcripts, subpoenaes, experts and attorneys must be paid. Further the hidden, "indirect cost" i.e., loss of employee production -- must also be paid.

It should be self evident that in complex construction cases the direct costs of an arbitration proceeding may equal or exceed the direct costs incurred in civil litigation.

CONCLUSION

Although arbitration is widely touted as a "speedy" and "inexpensive" method for dispute resolution, there are substantial risks to be assessed in considering whether to agree to employ the method. Not only can an arbitration proceeding in a complex case be lengthy, but also, delays can be encountered by way of disagreements over the arbitration panel or the filing of amended claims. The hearings themselves can be protracted if a party elects to submit volumes of evidence and many witnesses -- as noted previously, arbitrators tend to receive all evidence.

Parties to arbitrarion may save costs through avoiding the expense of depositions. However, they also face the risk of surprise attendent to prehearing discovery.

Arbitration litigants may also avoid the "cost" of transcribing the hearing at the risk of sacrificing the right of judicial review of an unfair or partial award.

Arbitration litigants may avoid some "cost" through hearings not encumbered by formal rules of evidence. However, they must also bear the "cost" of a decision predicated upon otherwise inadmissable evidence.

Finally, although arbitration litigants present their disputes to persons purportedly having "technical experience", frequently arbitrators have little or no "judicial experience." However, these judicially inexperienced arbitrators will often be called upon to decide difficult legal issues concerning the meaning of contracts, the standards of conduct, the reception of evidence and the scope or measure of damages. Moreover, their decisions will often be final.

In sum, the design professional should exercise great care before signing any contract requiring that disputes arising out of his or her work be submitted to arbitration. "Boiler plate" arbitration clauses present in many standard A.I.A. contracts should be carefully reviewed. See: e.g.: AIA Document B151, Owner-Architect Agreement; AIA Document A201, General Conditions for Owner/General Contractor Agreement; AIA Document B141, Owner-Architect Agreement. Indeed, thought should be given to either deleting such provisions from these contracts, or to very carefully limit the scope of these clauses to disputes arising during the course of construction or to disputes involving less than $10,000.

AN HISTORICAL PERSPECTIVE OF FAILURES OF CIVIL ENGINEERING WORKS

Neal FitzSimons, F.ASCE*

Abstract

The blight of failures has smitten civil engineering works since ancient times. An examination of historic cases reveals relationships to contemporary cases that are useful in guiding the development of policies and programs to reduce the incidence of failures. Considering human experience, knowledge and attitudes, a framework of teletic, technical and natural systems is useful for the study of these failures. Ignorance, incompetence, negligence and avarice are the four horsemen of the engineering apocalypse.

Introduction

Civil engineers have their roots in the earliest of civilizations. Even today there are vestiges of ancient roads, canals, dams, tunnels and other structures in Asia, Asia Minor, Africa, Europe, and in the western hemisphere. There is also evidence of a failure of performance of some of these works. Why have these failures occurred in the past? Why do they occur today? Are failures inevitable? Could the scope, severity and frequency of failures during any past era have been reduced? What actions have been taken by civil engineers in the past to avert failures? How successful were these actions?

To appreciate and evaluate any effort by our professional forebears to prevent the repetition of failures, it is necessary to understand their teletic (see Annex A) milieu as well as the technical and natural systems under which they operated.

The Tunnel of Eupalinus

Archaeologists have discovered an ancient (c. 500 B.C.) tunnel on the Greek island of Samos which has two distinct longitudinal bores connected by a perpendicular transverse bore. Apparently one of our early professional ancestors, Eupalinas of Megara, had erred. In this case, one can first ask: Was the engineer trying to extend the techniques of surveying beyond the norms of the time? (The tunnel was 3,300 feet; 1,007 m long.) If the answer is yes, then perhaps criticism is somewhat muted, however, there still is the question of a reckless attempt to extend beyond the norms. If the answer is no, and there was simply an error in the survey or in the construction, then another type of question arises: Who is responsible and to what degree?

*Principal, Engineering Counsel, Kensington, Maryland 20895

One can also ask other questions, such as: Was the overall project objective achieved despite the geometry of the final product? What remedial measures were proposed, e.g., continue both tunnels to the full extent or fair in a transition? Why were they rejected? Was the entire project abandoned because of the error? Did the responsible parties pay for the error? Did other tunnel engineers learn from the fiasco?

Seven hundred years later (c. 148 A.D.), a tunnel was built at Saldae (now Bougie, Algeria), this time by a Roman engineer, Nonius Datus. Just as with Eupalinus, there are two bores connected by a transverse bore. Here it cannot be said that the engineer was beyond the bounds of practice, but rather it appears that someone erred. In any event Datus wrote, "Thus I finished the work and the water passed through..." What punitive actions were taken by any of the teletic systems of his culture for this type of error? What actions should be taken today under similar technical circumstances?

Pont d'Avignon

Across a braid of the Rhone River near Avignon, there stands a series of four stone arches that terminate abruptly at the river. Eight centuries ago (c. 1180 A.D.), according to legend, a shepherd boy had from divine inspiration, conceived, designed and directed the construction of the greatest bridge in the world. After his death, he was buried in a chapel perched on one of the piers. Sadly, over the years, floods and ice jams came and swept away the fourteen river spans (some say fifteen and others sixteen), leaving what is seen today as an historical curiosity rather than an operating bridge. This magnificent ruin is the monument of Saint Benezet who is said to be the founder of the Fratres Pontifices (Brothers of the Bridge), who built many bridges in Europe until disbanded by Pope Clement IV. Did the shepherd cum engineer fail in his responsibility? How long should this bridge have performed its function before being labeled a failure? Even today, what engineering standards are there to guide the practitioner as to how long a structure should perform before it fails in service? How much maintenance and repair should an engineer expect when designing a bridge?

The Beauvais Cathedral and Saint Peter's Church

The urge to build grand structures for gold, glory or God perhaps began in earnest during the middle ages when the universal Christian church found the resources to inaugurate the great cathedral competition. On one hand, towers generated awe from the exterior while spacious domes created a sense of reverence from the interior. In both cases, structural failures were not uncommon. Perhaps the two greatest examples are the tower of the Cathedral at Beauvais, France and the dome of Saint Peter's Church in Rome, Italy.

Beginning about the IX Century, the towers of the great cathedrals reached higher and higher until today the spire of Ulm soars 529 feet (161 m) towards heaven. But the trail of failures from the 50 foot (15 m) bell tower in a parish church to the acme of spires in southern Germany was long and lugubrious. Perhaps the most dramatic of all was the collapse of the tower in the grand cathedral at Beauvais.

Its magnificence was unparalleled and it is said to have awed the most irreverent soul. The original construction of 1248 was halted by a major collapse in 1284, but after 40 years of remedial work, the hardy Beauvaisiens undertook to extend this great work. In 1569 a cross was placed on the newly completed tower 502 feet (153 m) above the pavement. It remained there until 1573 when the tower totally collapsed: both a technological and theological disaster. The faith of the masterbuilders in their ability to erect these great towers was shaken as was the faith of the common parishoner in the divine nature of the church itself. What is an acceptable rate of progress? Must the masterbuilder probe nature until a collapse occurs and then contemplate the reason? Or should the masterbuilder first contemplate the possible modes of collapse before probing? How often will teletic pressures overcome technical responsibility?

The present dome of Saint Peter's Church was indirectly responsible for a major milestone in structural engineering. When first designed and built circa 1548, its span of 138 feet (42 m) was slightly less than the Pantheon (124 A.D.) of Rome (142 feet: 43 m), but greater than Hagia Sophia (561 A.D.) in Constantinople (108 feet; 33 m). Unfortunately, at the time the mechanics of circumferential thrust were not well understood and in 1742 investigations uncovered severe cracking in the dome itself and in the supporting structure. The theological implications of a dome failure were so great that immediate action was required. Therefore a select group of three "mathematicians" were asked to determine the best method of achieving long term structural stability. This was perhaps the first well recorded incident of the application of theoretical mechanics to a project of major consequence. It was finally decided to install additional ties which proved successful and so it remains.

Here is found a classic case wherein the normal bounds of practice were traversed without appropriate attention given to monitoring the resultant structure. What provisions should have been made then and should be made now to properly monitor a structure that is advancing the "state of the art"? The real world is an excellent "laboratory" to probe interactions between natural and technical systems, but the probing must include intelligent collection, collation, analysis and dissemination of the data. Such probing (monitoring) should not only be able to reveal poor performance, if any, but also warn against disastrous performance.

Pont-y-Prydd

Near Cardiff in Wales lies the market town of Pontypridd, named after a remarkable bridge over the river Taff. Today is seen a beautiful single arch span of 140 feet (43 m) with spandrels pierced by symmetrical circular openings. The keystone is 36 feet (11 m) over the water and all its voussoirs are cut and fitted to a precision only a master mason can attain. For 235 years William Edwards (1719-1779) has been lauded for this masterwork, but in 1746 when he contracted to build a bridge over the Taff he was a little-known, 27-year-old farmer, mason and recently ordained minister.

His first attempt to cross the river had three 50 foot (15 m) arch spans with two piers. The completed structure was much admired, but did not survive the spring freshets of 1749. Perceiving scour as his natural adversary, he decided to rebuild without piers undaunted by the required span of 140 feet (43 m). At the time, there were perhaps two stone bridges in the world with such a span; the Claix bridge over the Drac near Grenoble, France and the Italian bridge at Trezzo over the Adda (destroyed in 1346). It should also be noted that the greatest Roman arch span, Ponte d'Augusto was about 105 feet (32 m) and of semi-circular geometry. Edwards proceeded with his second bridge and finally the time came to remove the timber centering, the temporary supports for the voussoirs. Unfortunately, at that time, the crown popped up and the haunches dropped down and the entire mass of masonry once again fell into the river.

Edwards called in John Smeaton (1724-1792), who suggested that the spandrels be perforated to reduce the dead weight of the soil over the haunches. Finally, using the revised design, the third bridge was completed in 1755 as it stands today. One post-script is the steep slopes in the final design made it so difficult for vehicles to cross that another bridge was built nearby as soon as it could be afforded. Ironically, the replacement had three masonry arch spans, similar to Edwards original design, but of slightly greater span. Unlike it, however, the later bridge had a level roadway. Another post-script concerns the use of perforated spandrels. Probably unknown to both Edwards and Smeaton was the fact that such an idea had been successfully incorporated into the Great Stone Bridge of China in the VII Century.

Pont-y-Pridd has many implications for the practitioner of today. First, should a specialist in masonry construction, unlearned even in the relative primitive hydraulics of his day, attempt to solve a problem larger than his ken? Second, should he attempt to go beyond the limits of conventional practice without fully using the best available knowledge? Third, should he compromise the function of a structure (carrying vehicles over a stream) by using a form which achieved strength but sacrificed geometry (a level road)?

Saint Chad, Shrewsbury

At the stroke of four o'clock on the morning of July 9, 1788, the great church of St. Chad collapsed. What was once the pride of Shrewsbury, England was now a pile of broken timbers and stone rubble.

A few weeks earlier, the Parish Vestry had called upon a young Scottish Architect/Engineer, Thomas Telford, to investigate the cracks in one of the four pillars which supported the bell tower. His examination revealed rotted roof timbers, shifted masonry and foundation movement caused by grave digging close to the footings. He recommended to the vestry an immediate major remedial program. Apparently, his proposal was rejected on the premise that the church had already stood four centuries, ergo it should stand up four more. Besides, the young Scot may have been more interested in gaining a major commission than in advising on a few minor repairs.

The vestry was wrong on both accounts and down came the church. It went down because members of the social system (a church vestry), distrustful of a member of the economic system (a professional practitioner) and ignorant of the technical system (a building) made a bad decision. The signals of the disaster were not clear enough to the vestry: the arguments of the architect were not persuasive enough and mutual respect was not present. Is this what happened at Selby in 1690 when the abbey church fell and later at Banbury, Chelmsford and Great Shefford? Is this what happened at Hereford in 1781 when the cathedral went from an awesome edifice of inspiration and beauty to a heap of ruins in minutes? Did Telford know the details about Hereford so that he could better persuade the vestry?

Ashtabula and Tay

In Northeast Ohio, a little stream called Ashtabula Creek runs into Lake Erie. It was a very unpretentious barrier for the Lake Shore & Southern Michigan Railroad. On December 29, 1876, the 154 foot (47 m) span all-iron Howe truss crossing collapsed while the 'Pacific Express' was traversing it and 80 people were killed (some accounts say 92). The structure had been in service 11 years. Many theories were advanced to account for the tragedy. The railroad's chief engineer, Charles Collins could offer no explanation and, after being savagely attacked in the press, committed suicide. His boss, Amasa Stone, stoutly defended the bridge's design, construction, operation and maintenance. Among the most prevalent reasons given was that somehow cast iron was the culprit.

Cast iron had been used in railroad bridges since 1829 when a few short-span (12.5 feet: 3.8 m) lenticular trusses of cast and wrought iron were introduced on the Stockton-Darlington Railroad in the Northeast of England. In the United States, stone arches and timber trusses gradually gave way to iron/timber combinations and finally to all-iron trusses after the Civil War (1861-1865) when iron became cheaper and railroad loads greater. Wrought iron was used for tension members and cast iron for compression members; steel was still generally unavailable.

According to Henry Tyrrell writing in 1911, "The failure . . . at Ashtabula . . . resulted in discarding cast iron entirely by railroad companies, and four or five years later it was also abandoned for highway bridges."

The Ashtabula case had many facets: a major political inquiry by the Ohio Legislature, massive newspaper coverage, and even a special session of the American Society of Civil Engineers. The irony is that bridge failures on American railroads became rampant after Ashtabula. Between 1878 and 1895, the Engineering News reported 502, or 34 per year. It's also ironic that even today it is difficult to find not only the statistics of failures such as these, but also a detailed analysis of their causes, physical or procedural.

Just three years after Ashtabula, the civil engineering profession was shocked to learn that the great Firth of Tay bridge had collapsed with the loss of over 200 souls. It had been in service two years, two months and two days. It also was constructed of wrought and cast iron because steel would not be permitted on British bridges until the year after completion of the bridge. The engineer, newly knighted Sir Thomas Bouch, was busily engaged in his latest project, the Firth of Forth crossing, when the disaster occurred.

Perhaps the first theory advanced for the failure was the easiest to understand--God disapproved of traveling on Sunday, ergo He made an example at Tay. Most other theories were concerned with materials and loads, particularly wind loads. An exhaustive government inquiry brought out many more facts than could be digested, but culpability seemed destined to settle on Sir Thomas who was relieved of his Forth responsibilities and who died shortly thereafter. Should he have used Smeaton's XVIII Century wind tables for guidance or conducted his own tests? How much peer review should he have requested on such an important project? Should he have permitted the use of cast iron at all? After all, there had been many warning signs as to its uncertain long-term strength properties. Were there operational factors that he had overlooked in his design or that he should have informed the owner about? Was his construction supervision adequate to insure a quality structure?

The replacement Firth of Tay Bridge has been standing for a century as has its southern neighbor, the Firth of Forth twin cantilever; both are of steel.

It might be noted that cast iron continued to be used in buildings until about 1913. Nine years earlier, after the collapse of the Darlington Hotel in New York, which killed 21 people, the Engineering Record editorialized, ". . . the hazardous characteristics of cast iron members of the class [columns] may easily be realized." How does a hazard become recognized as such by the design professions? If a hazard is recognized, what should be done about the structures already in place?

Notes on the Law and Engineering Failure

The greatest of Babylonian kings was Hammurabi, whose feats in war and peace were admired by all. His public works, canals, granaries and temples were huge enterprises. But, perhaps his greatest achievement of all was a written legal code. With reference to construction his ideas were quite simple: "If a contractor builds a house and it collapses killing its owner, the contractor will be killed. If the son of the owner is killed, then so will be the son of the contractor."

From this simple beginning, the laws associated with engineering and architectural failures grew more and more complex as did their means of adjudication. A milestone occurred in English common law on November 21, 1782 in the case of Folkes, Bart., v. Chadd and others. It seems that Chadd et al., as trustees of the harbor at Wells on the East coast of England, had obtained an injunction against Folkes to remove a protective dike around his low-lying pastures because it had been causing the silting up of the harbor. Folkes asked that the injunction be set aside and in the course of the trial, brought in John Smeaton, who testified that in his opinion the dike had not caused the "mischief."

Chadd's attorney asked that Smeaton's opinion be thrown out as hearsay, but the court ruled that, "Of this [matter], such men as Mr. Smeaton alone can judge. Therefore we are of opinion that his judgment, formed on facts, was very proper evidence."

A third note concerns the Napoleonic Code as promulgated in 1804. In it there seems to be an interesting interaction between the political and economic system: If there is a loss of serviceability in a constructed project within ten years of its completion because of foundation failure or from poor workmanship, the contractor and architect (luckily, not the engineer?) will be sent to prison.

The impact of Folkes v. Chadd is clearly evident on today's jurisprudence in which the costs of arbitration and litigation include not only legal fees and expenses, but also those for technical experts of all kinds. The impact of Hammurabi's or Napoleon's code on the reduction of failures then or now is questionable. How much does the threat of execution or imprisonment discourage poor practice? Are there fewer failures in areas under the Napoleonic code than under English common law? Has any meaningful study ever been done as to the interrelationship between laws, law enforcement and engineering failures? Have building laws been efficiently beneficial or even beneficial? Is law-based building inspection more effective than economic-based inspection?

Summary

This paper does not present conclusions, but rather asks questions about past failures of engineering, questions more teletic than technical. It implies that the procedural, i.e., human factors in these failures are at least as important as the physical, i.e., technical factors, if the evaluation of the failure is to be used to advance the art of engineering rather than just to collect compensation.

Benjamin Wright (1770-1842), the "Father of American Civil Engineering," chaired a committee in 1839 to found an "American Society of Civil Engineers." Its preamble stated, "It has been instituted for the collection and diffusion of professional knowledge . . . ." Is not performance data the most essential part of professional knowledge?

ANNEX A

Teletic Systems

A system is an orderly arrangement of units structured to facilitate the accomplishment of an objective. "Natural" systems are those which operate without human intervention in contrast to "Technical" systems, which are artificially built to operate under well-defined natural laws. "Teletic" systems are created by humans to achieve goals by the directed use of a society and its resources. Three teletic sub-systems are social, political and economic.

A social sub-system is concerned with the welfare of individual, families or cultural groups.

A political sub-system is concerned with the setting of societal goals and encouraging their achievement; providing and enforcing societal rules; and the acquisition and disposition of the means for performing these activities.

An economic sub-system is concerned with the utilization of resources to meet societal demands for goods, works and services.

Normally, individuals or organizations (social, political or economic) have roles and interactions with all systems and sub-systems at a given time in a given place. Further, these roles and interactions are usually changing with time, place and circumstances.

As a simple example, most members of ASCE play a role and interact with a family and church groups (social sub-system), as voters and taxpayers (political sub-system) and as workers and consumers (economic sub-system). When they drive their car, they are using many technical systems and, unfortunately, polluting (to some degree) the natural system.

FAILURE CLASSIFICATIONS

E. Chesson, Jr., F.ASCE*

Let's look back almost a hundred years to an article by George H. Thomson (M.Am.Soc.Eng.) published in ENGINEERING, Sept. 14, 1888, p. 252:

> "1. In considering mechanical structures, in either of two conditions in which they must present themselves to us, perfect or imperfect in fulfilling the purpose of their construction, we find it convenient and illustrative...to borrow the terms applicable to vital structures; to adopt as descriptive of a mechanism, and in elucidation of the method by which forces produce effects good or bad, the terms physiology and pathology.
>
> "Mechanical structural integrity, in securing a harmonious action of the aggregate parts concerned, forms no faint parallel with the human structure embodying vital completeness; we may call each a physiological success, while...a faulty or weak mechanism, and the resulting malfeasance in action, finds its counterpart in that state of the human system which is denominated pathological.
>
> "Thus, we can affirm the mechanical action of a bridge of good design, good manufacture, well maintained, and properly loaded, that it is physiological, or, we can affirm of the mechanical action of this same bridge if not properly cared for...or, if the bridge has been struck by a train, [that it is] pathological...
>
> "The statement that 251 railroad bridges have failed in the United States and Canada during the ten years ending December 31, 1887, from weakness, overwork, collision, etc., may be accepted as a mechanical pathological generalization...
>
> "Of the number 251, above quoted, 57 were knocked down by derailment, broken axles, etc.; 30 were 'square falls,' in 96 cases the real cause of failure was in a measure uncertain, five failures occurred during replacement or repairs.
>
> "Much difficulty is experienced in acquiring correct data, hence a classification of failures, cause, etc., is not easily arrived at."

---

*Professor, Department of Civil Engineering, University of Delaware, Newark, Delaware 19716.

Introduction

Failures, though unwelcome, continue to be among the most important lessons for engineers and architects, providing the bases for refinements in subsequent works. Failures have been classified by various schemes over the centuries, including the comparison above with living organisms! A few simplistic classifications are: ignorance, negligence, greed; improper design, construction, or maintenance; frightening, frustrating, fantastic, funny, or future; designer, builder, owner or user. Failures may be classified by location, project size, type of project, damage or injuries, components involved, or project stage at time of failure (age). Failure classifications may include considerations of external or "trigger" mechanisms: natural phenomena such as rain, snow, wind, earthquake or flooding, as well as unnatural phenomena such as auto or aircraft impact, overload, or blast. Almost any experienced design professional could add to this list of ways to classify project failures.

The recent trends in expanding litigation have led to more careful considerations of the causes and classifications of failures. It is important to mention that no project is likely to be totally perfect in design, construction, and operation. Careful checking and inspection may reveal minor errors of omission or commission on the part of one or more persons involved in even the best planned and executed project. But it is to be hoped that such minor errors will not combine to trigger a major failure. Of course, some major errors do occur which (if not caught soon enough) can lead to substantial loss of time, money, reputation and even life. High quality materials and performance are essential as well as careful checking and inspection. The possibility of failure is reduced when experienced, professionally curious, intelligent, and careful individuals are involved throughout a project.

In the 1888 article quoted above the author states that causes of failures were not easily determined a century ago. This is still true today, although to a lesser extent, thanks to better analytical procedures, improved knowledge and quality of materials, and more sophisticated methods of examining the factors involved. In selecting a classification system for engineering/architectural failures in our projects, it is desirable to know the reasons for seeking that information and the level of detail for which the causes should (or could) be determined. When we look at published reports of failures it is important to understand why that information was collected, the qualifications and experience of those collecting it, and the stage in the investigations at which the causal determinations were made. Rarely is this detail reported.

My own observations over the last two decades have been that most reports, even of an extensive statistical nature, do not really provide enough information to allow the reader to form precise conclusions. Instead, I believe that most such studies should be used only as indications of general trends rather than for specific statistical judgments. The conclusion that in a certain five-year period "45% of the failures were attributable to the designer" or that "28.3% of the cases were roof failures" might better be stated as "nearly half were thought to be..." or "about three out of ten..." Although this choice of wording appears less precise than calculated percentages, in reality it may actually be nearer to the truth. We are not usually able to pinpoint the exact causes of many failures, and as is shown later in this paper, the actual cause may not have been known when the classification of cause was made.

Need for a Comprehensive Classification System

A satisfactory classification system must address the needs of a variety of users:

Design professionals are interested in those failures which involve unsatisfactory behavior because of primarily technical and functional shortcomings. They must design in a reasonable time and at suitable profit a serviceable project that the owner can afford and that can actually be constructed by a qualified builder. Builders are more concerned with failures caused by workmanship, inferior materials, and temporary construction but must consider what can be accomplished at a price that will leave a profit despite the uncertainties of supplies and future costs of materials and labor. Owners may face problems because of poor performance by the designers or builders, and also because of operating personnel errors. Owners also require reasonable returns for capital investments and cannot afford unnecessary refinements and enhancements. Insurers are concerned with classifications of failures that permit them to set premiums and make equitable profits, whether the insurance is for designers, builders, or owners. Building officials are concerned about failures because of public safety and eventually because of property values and taxes. Rescue/disaster teams would likely classify failures in still other ways, because of a totally different mission. Time is vital in the saving of lives; protecting property and preserving evidence are of secondary importance.

It is evident that each party involved in a project has a somewhat different perspective. Since each involvement is unique and occurs at different times during the life of a project from inception to demolition, it is easy to understand why failure classifications in the past have stressed different aspects. Some classification systems have incorporated a degree of bias which was inherent in the data collected. Because of these differing perspectives classification systems have been established in ways that were not always sound nor logical. They may have introduced unintended distortions into the statistics that were subsequently developed from the original problem summaries, often years after the failures, by authors who had no first-hand involvement in the cases and who may have been unable to review the primary documents.

With the cooperation of both the Architectural and Engineering Performance Information Center at the University of Maryland and one of the nation's major insurers of designers, I had the opportunity to spend three months in early 1985 studying case histories on file with that insurer. It was decided that because of my background in civil engineering I would examine cases which were likely to involve the civil or structural engineer. It was also decided that only those cases would be included which had been settled (or for which all investigations were essentially complete) and which involved at least $30,000 liability on the part of the insured or insurer. In the course of the review, a few exceptions were made to these guidelines. The insurer did not deny me access to any cases on file. Because of that freedom of inspection, I was able to explore causes and trends, if any, with confidence.

In the case of this particular insurer, the coding of cause for a claim is usually done soon after the initiation of that claim. Insofar as I could determine, there is then no formal provision for subsequent review of that coding. This means that a failure may be labelled "brick veneer" when the problem is, for example, a leaking roof, a torsionally inadequate beam or lintel, or settlement of a foundation or grade beam. This initial coding is done by an insurance adjuster who may not necessarily be a design professional or experienced in construction. As a case unfolds, this insurance representative is involved with a number of other cases, arrangements, negotiations, and communications, and may not recall that the developing evidence in a specific case indicates that the initial coding should be corrected.

It is also important to note that during the development of a claim, conflicting professional opinions are often expressed as to the cause of a failure, sometimes by experts on the same side and frequently by those on different sides of the claim. Thus, even at the conclusion of a claim file, it may not be easy to assign with complete certainty the cause(s) of a failure. If a cause is to be established with high certainty, there should be a mechanism for review of a failure case file after the conclusion of the testimony by experts representing all sides. Preferably, this review should be conducted by a panel of three or more experienced professionals competent in the general field of the problem. The majority opinion of that panel should then replace the initial coding of the cause of failure for each claim. This process would be time-consuming and expensive but could overcome the present shortcomings in the data bases from companies which would likely have the most cases available for inclusion in statistical analyses.

During the three months of my association with the insurance underwriter, it was possible to review in depth a total of 154 cases from two categories: "wall/bearing" and "column or beam." In my opinion only 48 of these cases had been properly coded as to cause. If this sample is indicative of all data collected by that company, it means that only three out of ten cases being coded at initial stages by claims representatives will later be found to have been correctly identified as to cause. It is important to note that my own evaluations could be in error, and that in making my evaluations I had the advantage of being able to study the entire set of reports from experts and the court proceedings. However, if only three out of ten failure classifications are likely to be correct, considerable doubt is cast on the usefulness of the statistics which result, whether they are used by the insurance industry or by engineers, architects and others in the construction industry.

Some of these instances of improper coding can be explained by the fact that the coding form which had been in use since 1978 was revised in 1983, with some categories renumbered and others added. However, the computer program which was used for case retrieval and statistical studies had apparently not been reprogrammed to incorporate the changes. For example:

Classification	1978 Form	1983 Form
141	Bracing	Wall/Bearing
143	Welding	Slabs
162	Air Conditioning	Fire Protection

Clearly these (and similar) inconsistencies confuse and cloud the results. It should also be noted that the effects of changes in the coding classification as tabulated above will eventually disappear since the active data base in that particular firm is limited to five years. Thus, if the present classification form is used for some time, beginning in 1988 the categories will not reflect two meanings for data then being reported. Only the old statistical and data reports will contain errors from renumbering. Of course, there will still remain the unintentional but important errors that will result from incorrect classifications initiated before the true causes were determined. Many of the reclassifications I would have made were based on better information, my engineering background and substantial experience in studying and presenting failure case histories.

There were other difficulties with the insurance company forms in use at the time of my review. Few examples or explanations were included to aid the person assigning a classification to a failure. For example, there was a classification for "floors" and another for "slabs" but no guidelines as to when to use which term. How would a basement floor slab problem be coded?

Clearly, there should be a precise, carefully developed guide for the coder, whether an insurer or a designer. The coding form itself should be planned to reduce the likelihood of confusion and mistakes from careless reading. The form should be tested to determine if different evaluators would assign the same codes to a given and representative set of cases. Guidelines should be provided to explain how to differentiate among "water runoff," "roof problems," "waterproofing," and "other." ("Other" might apply when a roof leaks often because its drains have clogged due to insufficient inspection and debris removal.)

Failures involving formwork, bracing and guying appeared to be inadequately covered on the form in use at the time of my review. The only categories included that came close to covering these construction-related topics were: "floors," "slabs," "foundations," "trenching/shoring," "columns or beams," and "other." Yet none of these categories seems appropriate if the purpose is to isolate the causes of a large number of failures, construction injuries and deaths.

I want to emphasize that I am not expressing concern about one insurer and one series of forms. My concern is that what has been available to us for analysis of failures from all sources may have been clouded by some of the same kinds of problems that have come to light in my limited review. If my concerns have merit, then some of the past studies we are now using for guidance must either be taken lightly, discarded, or restudied.

In the future we may also need to incorporate into our categories failures resulting from the misuse of the now popular computer. For example:

(a) Improper data entry to a proper program
(b) Proper data entry into an inappropriate program
(c) Appropriate program with a previously unknown logic defect or "bug"
(d) Permutations of the above!

While I was reviewing cases for the insurer, I prepared at the same time a number of case history reports using the forms now available from the Architectural and Engineering Performance Information Center. It is my opinion that these forms and failure classifications are more specific and less likely to lead to classification errors, if completed by experienced construction professionals at the conclusion of a case. But if completed during initial stages of exploration there will still be the considerable chance that the designated "cause" will not be supported by the conclusions. I also feel that this set of forms should be modified in order to cover properly the range of cases that were revealed in the 154 files I examined in the spring of 1985.

Conclusions

The conclusions of this paper can be summarized as follows:

1. Any analysis of failure statistics from any source (or sources) may have unintended errors, depending on the reasons for and stages of collecting and classifying the original data.

2. It will be costly and time-consuming to review and revise a data bank to reflect the opinion(s) of one (or more) expert(s) after a review of the detailed case files. However, this may be the best way to obtain high quality data.

3. Care must be exercised in developing forms and instructions for collecting data and in the data management systems used. Otherwise confused codings are probable.

4. Finally, no system is likely to be without some inconsistencies; therefore, conclusions from the data should not be given more credence than is warranted by the quality of the information in the source files. Simply said, trends are more believable than percentages.

5. Some failures will never be satisfactorily classified.

An editorial in the August 22, 1985 Engineering News Record (p. 120) says:

> "The engineering ethic is to reduce complex, random-seeming natural events to mathematical analogs that predict those events and suggest reasons for their occurrence. The analytical models have obvious value. Real structures must be built in the dirty, incomprehensible real world. Without some way to reduce it to forms simple enough for a human brain to understand, nothing will every get built.
>
> "The trap in this is in taking the analogs too seriously, in thinking that they explain everything. For engineers, nothing

proves the fallibility of accepted models like a failure.

"When a group of dam experts met...recently, they rehashed the details of disasters dating back as far as 26 years and wound up admitting that they didn't know, and probably would never know, exactly what caused any of them...

"Who will ever know all the answers for sure? In order to set up the formulas you have to guess, using the best judgment that experience and limited information allow...

"But as the perpetually unexplained details of the spectacular failures remind us, any engineer's job description should list humility as an essential requirement."

Appendix A: AEPIC Performance Report (used by individuals)
Appendix B: AEPIC Document Citation (used by individuals)
Appendix C: AEPIC Classification System (used only by AEPIC)

ARCHITECTURE AND ENGINEERING PERFORMANCE INFORMATION CENTER PERFORMANCE REPORT

PART A: REPORTER, ACCESSIONS DATA

AEPIC, University of Maryland
3907 Metzerott Road
College Park, Maryland 20742
(301) 935-5544

A110 This form is intended for a full description of a specific Architecture or Engineering problem.* Because Performance Reports can focus on structural, electrical, mechanical, environmental or aesthetic functions, some categories of data may be inappropriate or unavailable for any one Report. Please add new categories if needed to better describe events reported.

A115 Form section numbers can be used as a format for word processor or computer drafting of Performance Reports. Enclose the printout, numbered to correspond to form sections, with a signed form.

A120 Investigative documents and other relevant reports, photographs, diagrams, or other materials should be enclosed with the signed form.

A121 Please note here documents, reports, photographs, etc., which are not enclosed, but which have been prepared in connection with the Reported Performance.

A122 ______
A123 ______

A130 **Questions:**

A131 Have you performed any services in connection with this Reported Performance?
______

A132 Are you willing to be contacted by *AEPIC* on this matter?
______

A133 May *AEPIC* refer other persons to you for information regarding this matter?
______

A140 **REPORTER** (A business card may be attached in lieu of writing name, etc.)

A150 Name ______
A155 Address ______

A160 Phone ______
A165 Occupation ______
A170 Registration/Licensure ______
A175 Professional Affiliation ______
A180 Present Position ______
A185 Organization ______

A190 Date this Report ______
Signature ______

*An *AEPIC* Document Citation form is used to report published and unpublished articles, reports, etc. for bibliographic data collection.

AEPIC use only

______
accession number

______
reporter code

acknowledge ______
log, inventory ______
cross reference ______
code ______
enter ______
file ______
update ______

Inventory-Repository Reference

R010 Report Form
R020 Test Reports
R030 Investigation Reports
R040 Drawings
R050 Photos
R060 Slides
R070 Contracts
R080 Litigation, arbitration docu.
R090 Judicial Opinions
R100 Settlement documents
R110 Published articles
R120 Keywords:

aepic pr 14 84 feb

PART B: PROJECT IDENTIFICATION & DESCRIPTION	Notes for Reporters
B200 TERMINOLOGY:	200 Technical or professional reference for terms used by Reporter; examples: CSI, ASTM, NSPE, AIA, NFPA, etc.
B210 PROJECT NAME, FUNCTION	210 Full project name, number designation, if any. Specify use if not obvious from name.
B220 LOCATION	220 Address for buildings; for other projects, nearest place, milestone, river, topographic feature. Give State if U.S.A.
B230 OVERALL PROJECT SYSTEMS, MATERIALS	230 Describe project structural systems, construction type, structural, finish materials.
B240 OVERALL PROJECT DIMENSIONS	240 Describe size by dimension appropriate to project: length, width, stories, span, square feet, meters, height, etc. Use approximate measure if no exact size known.
B250 PROJECT DATES, COST	250, 255 Date begun, completed; original cost; changes, reason, dates. Use range if no exact date. Costs as known; approximate or estimate if unknown.
B255 ALTERATIONS DURING OR SINCE CONSTRUCTION	
B260 COMPONENT(S), ELEMENT(S) INVOLVED, MATERIALS	260 Describe part, system, subsystem, component, space or area, of project directly involved in problem events.
B270 COMPONENT, ELEMENT DIMENSIONS	270 See 240. Dimensions may be noted on sketch in Part D.
B280 CONDITIONS, AGENTS, CATALYSTS	280 Specify loads, pressures, forces, temperature and weather conditions. Note accumulations, impact, vibrations, etc.; designate factors as typical, unusual or extreme.
B285 CONTRACTS, CODES, LAWS	285 Specify contracts, codes, laws relevant to problem. Note whether allegedly violated, possibly applicable, violation established, etc.
B290 EQUIPMENT, PRODUCTS, TOOLS	290 List proprietary or generic type of product, transport, erection or fabrication equipment, tools involved. Listed items need not be the cause of the problem, but can be part of a remedy or repair procedure. Use Part C to describe use or function of listed items.
B295 SPECIAL FACTORS	295 Unspecified data: Notes of experimental nature of systems, materials, components used, nonstandard applications, unusual conditions, etc.
	aepic pr24 84 feb

PART C: PERFORMANCE EVENTS, ANALYSIS, CONCLUSIONS AND RECOMMENDATIONS	
C300 DESCRIPTION OF EVENTS:	**Reporter's checklist for Part C** (Data of special importance to *AEPIC* are **in bold type**.) ____ Date, timespan of reported events; year, month or season if no date known. ____ Stage or phase of Project; construction, survey, design, occupation, alteration, etc. ____ Signs, conditions, precursors, warning signs, accompanying factors. ____ Observed events, dates and timing, discovery, diagnosis, **initial remedial measure.** ____ Role or title of persons involved, in whose employ; "involved" includes all those taking part or affected, including the discoverer of errors, agent of repair, or victim. ____ Results: injuries, deaths, economic losses, time out of service, demolition, etc. ____ **Replacement, repair, reconstruction, problem solution.** ____ Estimate of actual total losses due to problem malfunction, failure. ____ Progress of legal proceedings, if any, or other dispute resolution process. ____ Outcome; settlement or allocation of duties, repair, payments, damages.
C310 ANALYSIS:	____ Analyses, investigations performed; by whom: reporter, investigator, government official, etc.
C320 CONCLUSIONS:	____ Apparent or established major cause or last agent, factor or error. ____ Apparent or established contributing agents, factors or errors; specific acts or omissions, missing information, unknown facts, miscommunication, noncommunication.
C330 RECOMMENDATIONS:	____ Overall recommendations. **Proposed act or conduct which would have avoided problem or lessened severity.** ____ **Proposed changes in practice** or **procedure to incorporate lessons in quality assurance industry-wide.** aepic pr34 84 feb

PART D: SKETCHES, KEYWORDS, COMMENTS

D410 Caption:

D420 Scale:

D430 Keywords: (List the most descriptive words related to the event.)

D440 Comments, postscripts:

aepic pr44 84 feb

ARCHITECTURE AND ENGINEERING PERFORMANCE INFORMATION CENTER DOCUMENT CITATION

DOCUMENT: P100 Published ________ Unpublished ________ material*
P110 TITLE ____________________

P200 AUTHOR(S) ____________________

P300 SOURCE, PUBLISHER ____________________

P400 PUBLISHER CITY, STATE ____________________

P500 PERIODICALS:
P510 DATE ____________ ***P520*** VOL ______ ***P530*** NO ______
P540 PAGES ______ to ______
P600 BOOKS, PAMPHLETS, SERIALS:
P610 DATE ____________________
P700 SERIES TITLE ____________________

P800 EVENT DATE or TIMESPAN: ____________________

P820 ABSTRACT (Contents, coverage, conclusions; note if photocopy enclosed.)

P850 KEYWORDS (List the most descriptive words related to the event.)

P900 **REPORTER** (A business card may be attached in lieu of writing name, etc.)
P950 Name ____________________
P955 Address ____________________

P960 Phone ____________________
P965 Occupation ____________________
P970 Registration/Licensure ____________________
P975 Professional Affiliation ____________________
P980 Present Position ____________________
P985 Organization ____________________

P990 Date this Report ____________________

Signature ____________________

*An *AEPIC* Performance Report form is used to describe a specific Architecture or Engineering problem for case report data collection.

AEPIC, University of Maryland
3907 Metzerott Road
College Park, Maryland 20742
(301) 935-5544

AEPIC use only

accession number

reference case number

reporter code

Document Source: DS __ __ __
Document Type: DT __
Document Class: DC __ __ __
Quick Codes:
__ __ __ __ __ CSI ref. no.
__ __ __ __ __ CSI ref. no.
__ __ __ __ __ CSI ref. no.
QL __ __ LOCATION
QP __ __ PROBLEM
QU __ __ OVERALL PROJECT USE
QS __ __ PROJECT SIZE
QC __ __ COMPONENT INVOLVED
QM __ __ MATERIALS/TYPE
QY __ __ YEAR
QT __ __ TIMESPAN/MONTH
QD __ __ DEATHS
QI __ __ INJURIES
QR __ __ CATALYST/RESULT
QO __ __ OUTCOME/COST
QF __ __ CENTRAL FACTOR

Keywords

aepic dcll 84 feb

AEPIC QUICK CODES
Edition 0984

Architecture and Engineering
Performance Information Center
AEPIC, University of Maryland
3907 Metzerott Road
College Park, Maryland 20742
(301) 935-5544

ALL FIELDS: xx = NO DATA
00 = NONE

DOCUMENT CITATION CODES

DOCUMENT SOURCE

100 AEPIC
200 Professional organizations
240 Architectural firms
280 Engineering firms
300 Consultants, multidiscipline
320 Testing laboratories
340 Law firms
400 Educational institutions
450 Media
500 Construction industry
550 Technical associations
600 State governments
670 Local governments
700 Foreign governments
751 Foreign organizations
800 U.S. Congress
- 801 House Committees
- 802 Senate Committees
- 803 Library of Congress
- 804 General Accounting Office
- 805 Budget Office

850 Executive Branch
- 851 Bureau of Census
- 852 Consumer Product Safety Comm.
- 853 Coast Guard
- 854 Commerce Department
- 855 Dept. of Energy
- 856 Dept. of Transportation
- 857 Dept. of Defense
- 858 Environmental Protection Agency
- 859 Executive Office
- 860 Federal Trade Commission
- 861 General Services Administration
- 862 Geological Survey
- 863 Housing, Urban Development
- 864 Interior Department
- 865 International Trade Commission
- 866 Justice Department
- 867 Nat. Inst. Occup. Safety & Health
- 868 Occup. Safety & Health Admin.
- 869 Veterans Administration

900 U.S. Judicial Branch
- 901 Federal Courts
- 950 State Courts

DOCUMENT TYPE

A Article, published
B Bibliography, search, index
C Conference report, proceeding
D Directory, dictionary
E Environmental analyses, filings
F Financial report, fiscal matter
G Guide, handbook
H Hearing, history
I Investigation, inspection, research
J Journal, collected case histories
K Contract, agreement
L Law, legislative document
M Major dossier, case, claim document
N (reserved)
O Opinion, case law, decision, ruling
P Policy statement, position paper
Q (reserved)
R Regulation, rule
S Specification, Code
T Trial, litigation, brief, memorandum
U Textbook
V Visual material, drawing, photo
W Working paper, analysis
X (reserved)
Y Yearbook, map, model, exhibit
Z Not elsewhere classified

DOCUMENT CLASS

Up to three digits, each representing a topic, in order of emphasis.

0 Unknown
1 Design
2 Construction
3 Testing, Research
4 Structure
5 Materials, Products
6 Information science, Computers
7 Legal matters
8 Insurance, Risk Management
9 Finance, Statistics

CSI Reference Code:
CSI _ _ _ _ _
CSS _ _ _ _ _
CST _ _ _ _ _

DIV 1–GENERAL REQUIREMENTS
01010 SUMMARY OF WORK
01100 ALTERNATIVES
01200 PROGRESS AND PAYMENT
01300 SUBMITTALS
01400 TESTINIG LABORATORY SERVICES
01500 TEMP. FACILITIES, CONTROLS
01600 MATERIAL AND EQUIPMENT
01700 PROJECT CLOSEOUT

DIV 2–SITE WORK
02010 SUBSURFACE EXPLORATION
02100 CLEARING
02110 DEMOLITION
02200 EARTHWORK
02250 SOIL POISONING
02300 PILE FOUNDATIONS
02350 CAISSONS
02400 SHEETING, SHORING, BRACING
02500 SITE DRAINAGE
02550 SITE UTILITIES
02600 PAVEMENTS AND WALKS
02700 SITE IMPROVEMENTS
02800 LANDSCAPING
02850 RAILROAD WORK
02900 MARINE WORK
02950 TUNNELING

DIV 3–CONCRETE
03100 CONCRETE FORMWORK
03150 EXPANSION, CONTRACTION JOINTS
03200 CONCRETE REINFORCEMENT
03300 CAST-IN-PLACE CONCRETE
03350 SPECIALLY FINISHED CONCRETE
03400 PRECAST CONCRETE
03500 CEMENTITIOUS DECKS

DIV 4–MASONRY
04100 MORTAR
04150 ACCESSORIES
04200 UNIT MASONRY
04400 STONE
04500 MASONRY RESTORATION, CLEANING
04550 REFRACTORIES

DIV 5–METALS
05100 STRUCTURAL METAL
05200 STEEL JOISTS
05300 METAL DECKING
05400 LIGHTGUAGE FRAMING
05500 MISCELLANEOUS METAL
05700 ORNAMENTAL METAL

DIV 6–WOOD AND PLASTICS
06100 ROUGH CARPENTRY
06130 HEAVY TIMBER CONSTRUCTION
06150 TRESTLES
06170 PREFABRICATED STRUCTURAL WOOD
06200 FINISH CARPENTRY
06300 WOOD TREATMENT
06400 ARCHITECTURAL WOODWORK
06500 PREFAB. STRUCTURAL PLASTICS
06600 PLASTIC FABRICATIONS

DIV 7–THERMAL, MOISTURE PROTECTION
07100 WATERPROOFING
07150 DAMPPROOFING
07200 INSULATION
07300 SHINGLES AND ROOFING TILES
07400 PREFORMED ROOFING AND SIDING
07500 MEMBRANE ROOFING
07600 FLASHING AND SHEET METAL
07800 ROOF ACCESSORIES
07900 CAULKING AND SEALANTS

DIV 8–DOORS AND WINDOWS
08100 METAL DOORS AND FRAMES
08200 WOOD AND PLASTIC DOORS
08300 SPECIAL DOORS
08500 METAL WINDOWS
08600 WOOD AND PLASTIC WINDOWS
08700 HARDWARE AND SPECIALTIES
08800 GLAZING
08900 CURTAINWALL SYSTEM
08950 STOREFRONT SYSTEM

DIV 9–FINISHES
09100 LATH AND PLASTER
09250 GYPSUM DRYWALL
09300 TILE
09400 TERRAZZO
09450 VENEER STONE
09500 ACOUSTICAL TREATMENT
09550 WOOD FLOORING
09650 RESILIENT FLOORING
09680 CARPETING
09700 SPECIAL FLOORING
09800 SPECIAL COATINGS
09900 PAINTING
09950 WALL COVERING

DIV 10–SPECIALTIES
10100 CHALKBOARDS, TACKBOARDS
10130 CHUTES
10150 COMPARTMENTS, CUBICLES
10230 DISAPPEARING STAIRS
10240 DOCK FACILITIES
10250 FIREFIGHTING DEVICES
10300 FIREPLACES
10350 FLAGPOLES
10400 IDENTIFYING DEVICES
10500 LOCKERS
10530 PLASTIC SPECIALTIES
10550 POSTAL SPECIALTIES
10600 PARTITIONS
10650 SCALES
10670 STORAGE SHELVING
10700 SUN CONTROL DEVICES (EXT)
10750 TELEPHONE ENCLOSURES
10800 TOILET, BATH ACCESSORIES
10900 WARDROBE SPECIALTIES
10950 WASTE DISPOSAL UNITS

DIV 11–EQUIPMENT
11100 BANK EQUIPMENT
11150 COMMERCIAL EQUIPMENT
11170 CHECKROOM EQUIPMENT
11180 DARKROOM EQUIPMENT
11200 ECCLESIASTICAL EQUIPMENT
11300 EDUCATIONAL EQUIPMENT
11400 FOOD SERVICE EQUIPMENT
11500 ATHLETIC EQUIPMENT
11550 INDUSTRIAL EQUIPMENT
11600 LABORATORY EQUIPMENT
11630 LAUNDRY EQUIPMENT
11650 LIBRARY EQUIPMENT
11700 MEDICAL EQUIPMENT
11800 MORTUARY EQUIPMENT
11830 MUSICAL EQUIPMENT
11850 PARKING EQUIPMENT
11880 PRISON EQUIPMENT
11900 RESIDENTIAL EQUIPMENT
11960 SHIPYARD EQUIPMENT
11970 THEATRE EQUIPMENT

DIV 12–FURNISHINGS
12100 ARTWORK
12200 BLINDS AND SHADES
12300 CABINETS AND FIXTURES
12500 DRAPERY AND CURTAINS
12600 FURNITURE
12670 RUGS AND MATS
12700 SEATING

DIV 13–SPECIAL CONSTRUCTION
13010 AIR SUPPORTED STRUCTURES
13050 ACCESS FLOORING
13100 AUDIOMETRIC ROOM
13250 CLEAN ROOM
13300 GREENHOUSE
13350 HYPERBARIC ROOM
13400 INCINERATORS
13440 INSTRUMENTATION
13450 INSULATED ROOM
13500 INTEGRATED CEILING
13540 NUCLEAR REACTORS
13550 OBSERVATORY
13650 PREFAB STRUCTURES
13700 RADIATION PROTECTION
13750 CHIMNEYS
13770 SOUND ISOLATION
13800 STORAGE VAULTS
13850 SWIMMING POOLS

DIV 14–CONVEYING SYSTEMS
14100 DUMBWAITERS
14200 ELEVATORS
14300 HOISTS AND CRANES
14400 LIFTS
14500 MAT'L HANDLING SYSTEM
14600 MOVING STAIRS, WALKS
14700 PNEUMATIC TUBE SYSTEM

DIV 15–MECHANICAL
15010 GENERAL PROVISIONS
15100 BASIC MAT'LS METHODS
15180 INSULATION
15200 WATER SUPPLY, TREATMT
15300 WASTEWATER DISP/TRTMT
15400 PLUMBING
15550 FIRE PROTECTION
15600 POWER/HEAT GENERATION
15650 REFRIGERATION
15700 LIQUID HEAT DISTRIB.
15800 AIR DISTRIBUTION
15900 CONTROLS, INSTRUMNTN.

DIV 16–ELECTRICAL
16010 GENERAL PROVISIONS
16100 BASIC MAT'LS, METHODS
16200 POWER GENERATION
16300 OUTSIDE POWER TRANS-MISSION, DISTRIBUTION
16400 SERVICE, DISTRIBUTION
16500 LIGHTING
16600 SPECIAL SYSTEMS
16700 COMMUNICATIONS
16850 HEATING AND COOLING
16900 CONTROLS, INSTRUMNTN.

AEP C QUICK CODES - page 2 (edition 0984)

AFGHANISTAN	AF
ALBANIA	AL
ALGERIA	AG
AMERICAN SAMOA	AQ
ANDORRA	AN
ANGOLA	AO
ANGUILLA	AV
ANTARCTICA	AY
ANTIGUA & BARBUDA	AC
ARGENTINA	AR
ASHMORE & CARTIER IS	AT
AUSTRALIA	AS
AUSTRIA	AU
BAHAMAS	BF
BAHRAIN	BA
BAKER I	FQ
BANGLADESH	BG
BARBADOS	BB
BASSAS DE INDIA	BS
BELGIUM	BE
BELIZE	BH
BENIN	BN
BERMUDA	BD
BHUTAN	BT
BOLIVIA	BL
BOTSWANA	BC
BOUVET I	BV
BRAZIL	BR
BRIT INDIAN OCEAN TERR	IO
BRIT VIRGIN IS	VI
BRUNEI	BX
BULGARIA	BU
BURMA	BM
BURUNDI	BY
CAMEROON	CM
CANADA	CA
CAPE VERDE	CV
CAYMAN IS	CJ
CENTRAL AFRICAN REP	CT
CHAD	CD
CHILE	CI
CHINA	CH
CHRISTMAS I	KT
CLIPPERTON I	IP
COCOS (KEELING) IS	CK
COLOMBIA	CO
COMOROS	CN
CONGO	CF
COOK IS	CW
CORAL SEA IS	CR
COSTA RICA	CS
CUBA	CU
CYPRUS	CY
CZECHOSLOVAKIA	CZ
DENMARK	DA
DJIBOUTI	DJ
DOMINICA	DO
DOMINICAN REP	DR
ECUADOR	EC
EGYPT	EG
EL SALVADOR	ES

EQUATORIAL GUINEA	EK
ETHIOPIA	ET
EUROPA I	EU
FAROE IS	FO
FALKLAND IS	FA
FIJI	FJ
FINLAND	FI
FRANCE	FR
FRENCH GUIANA	FG
FRENCH POLYNESIA	FP
FRENCH S & ANTARC LANDS	FS
GABON	GB
GAMBIA	GA
GAZA STRIP	GZ
GERMAN DEM REP	GC
GERMANY, BERLIN	BZ
GERMANY, FED REP	GE
GHANA	GH
GIBRALTAR	GI
GLORIOSO IS	GO
GREECE	GR
GREENLAND	GL
GRENADA	GJ
GUADALOUPE	GP
GUAM	GQ
GUATEMALA	GT
GUERNSEY	GK
GUINEA	GV
GUINEA-BISSAU	PU
GUYANA	GY
HAITI	HA
HEARD I & MCDONALD I	HM
HONDURAS	HO
HONG KONG	HK
HOWLAND I	HQ
HUNGARY	HU
ICELAND	IC
INDIA	IN
INDONESIA	ID
IRAN	IR
IRAQ	IZ
IRAQ-SAUDI AR NEUT ZN	IY
IRELAND	EI
ISRAEL	IS
ITALY	IT
IVORY OCAST	IV
JAMAICA	JM
JAN MAYEN	JN
JAPAN	JA
JARVIS I	DQ
JERSEY	JE
JOHNSTON ATOLL	JQ
JORDAN	JO
JUAN DE NOVA I	JU
KAMPUCHIA	CB
KENYA	KE
KINGMAN REEF	KQ
KIRIBATI	KR
KOREA DEM PEOP REP	KN
KOREA, REP	KS
KUWAIT	KU

LAOS	LA
LEBANON	LE
LESOTHO	LT
LIBERIA	LI
LIBYA	LY
LIIECHTENSTEIN	LS
LUXEMBOURG	LU
MACAU	MC
MADAGASCAR	MA
MALAWI	MI
MALAYSIA	MY
MALDIVES	MV
MALI	ML
MALTA	MT
MAN, ISLE OF	IM
MARTINIQUE	MB
MAURITANIA	MR
MAURITIUS	MP
MAYOTTE	MF
MEXICO	MX
MIDWAY IS	MQ
MONACO	MN
MONGOLIA	MG
MONTSERRAT	MH
MOROCCO	MO
MOZAMBIQUE	MZ
NAMIBIA	WA
NAURU	NR
NAVASSA I	BQ
NEPAL	NP
NETHERLANDS	NL
NETHERLANDS ANTILLES	NA
NEW CALEDONIA	NC
NEW ZEALAND	NZ
NICARAGUA	NU
NIGER	NG
NIGERIA	NI
NIUE	NE
NORFOLK I	NF
NORTHERN MARIANA IS	CQ
NORWAY	NO
OMAN	MU
PAKISTAN	PK
PALMYRA ATOLL	LQ
PANAMA	PM
PAPUA NEW GUINEA	PP
PARACEL IS	PF
PARAGUAY	PA
PERU	PE
PHILIPPINES	RP
PITCAIRN IS	PC
POLAND	PL
PORTUGAL	PO
PUERTO RICO	RQ
QATAR	QA
REUNION	RE
ROMANIA	RO
RWANDA	RW
ST CHRISTOPHER & NEVIS	SC
ST HELENA	SH
ST LUCIA	ST

Field ab: LOCATION (QL __ __) Country

Each country will use a list of standard number or letter abbreviations for provinces or regional divisions.

ST PIERRE & MIQUELON	SB
ST VINCENT & GRENADINES	VC
SAN MARINO	SM
SAO TOME & PRINCIPE	TP
SAUDI ARABIA	SA
SENEGAL	SG
SEYCHELLES	SE
SIERRA LEONE	SL
SINGAPORE	SN
SOLOMON IS	BP
SOMALIA	SO
SOUTH AFRICA	SF
SPAIN	SP
SPRATLY IS	PG
SRI LANKA	CE
SUDAN	SU
SURINAME	NS
SVALBARD	SV
SWAZILAND	WZ
SWEDEN	SW
SWITZERLAND	SZ
SYRIA	SY
TAIWAN	TW
TANZANIA, UN REP	TZ
THAILAND	TH
TOGO	TO
TOKELAY	TL
TONGA	TN
TRINIDAD & TOBAGO	TD
TROMELIN I	TE
TRUST TERR OF PACIFIC IS	NQ
TUNISIA	TS
TURKEKY	TU
TURKS & CAICOS IS	TK
TUVALU	TV
UGANDA	UG
UNION OF SOV SOC REPS	UR
UNITED ARAB EMIRATES	TC
UNITED KINGDOM	UK
UNITED STATES	US
UPPER VOLTA	UV
URUGUAY	UY
VANUATU (NEW HEBRIDES)	NH
VATICAN CITY	VT
VENEZUELA	VE
VIETNAM	VM
VIRGIN IS OF US	VQ
WAKE I	WQ
WALLIS & FORTUNA	WF
WEST BANK	WE
WESTERN SAHARA	WI
WESTERN SAMOA	WS
YEMEN (ADEN)	YS
YEMEN (SANAA)	YE
YUGOSLAVIA	YO
ZAIRE	CG
ZAMBIA	ZA
ZIMBABWE	ZI

AEPIC QUICK CODES - page 3 (edition 0984)

Field ba: SUBLOCATION (LQ __ __)

State, Region

Each country will use a list of standard number or letter abbreviations for provinces or regional divisions.

State Codes

01 AL Alabama
02 AS Alaska
03 AZ Arizona
04 AK Arkansas
05 CA California
06 CO Colorado
07 CN Connecticut
08 DE Delaware
09 DC Dist. of Col.
10 FL Florida
11 GA Georgia
12 HA Hawaii
13 ID Idaho
14 IL Illinois
15 IN Indiana
16 IA Iowa
17 KN Kansas
18 KY Kentucky
19 LA Louisiana
20 ME Maine
21 MD Maryland
22 MA Massachusetts
23 MI Michigan
24 MN Minnesota
25 MS Mississippi
26 MO Missouri
27 MT Montana
28 NB Nebraska
29 NV Nevada
30 NH New Hampshire
31 NJ New Jersey
32 NM New Mexico
33 NY New York
34 NC North Carolina
35 ND North Dakota
36 OH Ohio
37 OK Oklahoma
38 OR Oregon
39 PA Pennsylvania
40 RI Rhode Island
41 SC South Carolina
42 SD South Dakota
43 TN Tennessee
44 TX Texas
45 UT Utah
46 VT Vermont
47 VA Virginia
48 WA Washington
49 WV West Virginia
50 WS Wisconsin
51 WY Wyoming

52 Multiple states

Field cd: PROBLEM (QP __ __)

1st digit

1 Thermal, moisture penetration
2 Interior, spatial dysfunction
3 Structural integrity compromise
4 Movement, deflection
5 Electrical malfunction
6 Mechanical malfunction
7 Fire, disaster
8 Environmental discomfort, hazard
9 Acoustic, visual, aesthetic dysfunction

2nd digit

1 Human act, omission
2 Materials, maintenance failure

3 Human and materials, maintenance factors

4 External events, natural conditions

5 Conditions, events plus human factor
6 Conditions, events & materials, maintenance

7 Design, practice error

8 Design, practice error plus human factor
9 Design, practice error plus materials, maintenance
0 Design, practice error plus conditions, events

Field ef: OVERALL PROJECT USE (QU __ __)

Structures

01 Special ______________
02 Airport, navaid, fueling
03 Airfield, paving
04 Bin, silo
05 Bridge, trestle, viaduct
06 Cableway
07 Communications dish
08 Causeway
09 Containment vessel
10 Culvert
11 Dam
12 Derrick
13 Dike, levee
14 Dock, wharf
15 Drainage works
16 Electricity generation
17 Embankment
18 Excavation
19 Formwork, shoring
20 Foundation structure
21 Harbor, jetty, pier
22 Harbor, terminal
23 Highway, road
24 Hoist, crane
25 Hydraulic structure
26 Incinerator
27 Irrigation system
28 Lighthouse
29 Monument
30 Offshore structure
31 Park
32 Parking area
33 Pipeway
34 Railway
35 Refinery
36 Retaining wall
37 Scaffolding
38 Seawall, breakwater
39 Sewage processing
40 Stack, chimney
41 Subaqueous structure
42 Swimming pool
43 Tank
44 Tower, cooling
45 Tower, freestanding
46 Tower, guyed
47 Tunnel, subway
48 Wall, barrier
49 Water tower
50 Water processing
51 Waterway

Buildings

55 Airport terminal, hangar
56 Airport freight, storage
57 Arena, convention hall
58 Auditorium, theatre
59 Chemical plant
60 Church, chapel
61 Commercial, retail
62 Computer facility
63 Courthouse
64 Dormitory
65 Educational facility
66 Field house, gymnasium
67 Food establishment
68 Freight terminal
69 Garage
70 Hotel, motel
71 Housing, highrise
72 Housing, multiple unit, low-rise
73 Housing, detached
74 Industrial, heavy
75 Industrial, light
76 Laboratory, research
77 Library
78 Medical facility
79 Museum, gallery
80 Nuclear facility
81 Office building
82 Parking deck
83 Postal facility
84 Public building
85 Prison, correctional
86 Recreation facility
87 Refrigeration facility
88 Service station
89 Shelter
90 Shopping center
91 Stadium
92 Transportation terminal
93 Warehouse

Field gh: OVERALL PROJECT SIZE (QS __ __)

1st digit

1 Building
2 Nonbuilding
3 Multistructure complex
4 Open span structure
5 Special

2nd digit

Nonbuildings

1 Small project, light construction
2 Medium project
3 large project, heavy construction

Buildings

1 1 - 3 stories
2 4 - 9 stories
3 10+ stories

4 Multiple level, midrise at maximum
5 Multilevel, highrise at maximum
6 Multilevel tall buildings

7 Short open span
8 Medium open span
9 Longspan structure

AEPIC QUICK CODES - page 4 (edition 0984)

Field ij: COMPONENT, ELEMENT INVOLVED (QC __ __)

10 Site, subsurface
11 Excavation, grading, compaction
12 Sheeting, bracing
13 Piles, caissons
14 Drainage
15 Bedding
16 Tunnel lining
17 Retaining wall
18 Dam, cofferdam
19 Special

20 Substructure, foundation
21 Footings, line
22 Footings, mat
23 Footings, column
24 Pier
25 Wall
26 Buttress
27 Pile cap
28 Abutment
29 Special

30 Structure
31 Vertical system
32 Horizontal system
33 Continuous structure
34 Anchorage
35 Connection
36 Joint
37 Arch, shell
38 Suspension, membrane
39 Special

40 Exterior, envelope
41 Window
42 Door
43 Roof
44 Wall panel
45 Insulation
46 Waterproofing
47 Flashing
48 Caulk, sealant
49 Special

50 Interior
51 Wall
52 Floor
53 Ceiling
54 Horizontal circulation
55 Vertical circulation
56 Core
57 Spaces
58 Surfaces
59 Special

60 Temporary construction
61 Bracing
62 Shoring
63 Formwork
64 Scaffolding
65
66
67
68
69

70 Mechanical Systems
71 Cooling
72 Heating
73 Ventilation
74 Plumbing
75 Lighting, power
76 Transport
77 Hazard detection, protection
78 Emergency power, supply
79 Special

80 Paving, landscape
81 Roadway
82 Bridgedeck
83 Channel lining
84 Trenching
85 Drainage
86
87
88
89

90 Special construction
91 Marine installation
92 Oil, gas, other installation
93 Tower, stack, chimney
94 Water containment
95 Toxic materials handling
96 Low voltage electricity
97 High voltage electricity
98
99

Field kl: COMPONENT MATERIALS, AGE (QM __ __)

1st digit

1 Steel, steel components
2 Other metals, alloys
3 Cement, mortar, masonry
4 Concrete, mineral aggregates
5 Glass, tile, ceramics
6 Bitumin, asphalt, paint
7 Coatings, sealants, plastic, rubber, membrane
8 Building stone, earthworks
9 Wood, interior coverings

2nd digit

Years in service, 0 = construction period
x = unknown

Field mn: YEAR of PROBLEM (QY __ __)

Field op: TIMESPAN/MONTH (QT __ __)

1 January
2 February
3 March
4 April
5 May
6 June
7 July
8 August
9 September
0 October
1 November
2 December

3 Spring, month unspecified
4 Summer
5 Fall
6 Winter

7 No date, period of time under six months
8 No date, period six to twelve months
9 No date, period twelve to eighteen months
0 No date, period eighteen to twenty-four months

4 to 99 Months' duration of problem

Field qr: DEATHS (QD __ __)

Code number, 01 to 99. Code over 99 as 99.

Field st: INJURIES (QI __ __)

Code number, 01 to 99. Code over 99 as 99.

Field uv: CATALYST/RESULT (QR __ __)

1st digit

0 Loads
1 Cold, freeze-thaw
2 Heat, expansion
3 Wind
4 Water, moisture
5 Vibration, sound
6 Impact
7 Equipment, operation failure
8 Geotechnical movement
9 Fire, explosion

2nd digit

1 minor, surface, aesthetic or functional problem
2 crack, warning, minor mechanical problem
3 weather integrity impaired, penetration damage
4 minor deformation, acoustic or temperature problem
5 major deformation, large water entry
6 sinking, settling, major mechanical malfunction
7 minor, partial collapse, destruction
8 major collapse, destruction
9 Special ____

Field wx: OUTCOME/COST (QO __ __)

1st digit	2nd digit
1 Notice of problem	1 under K 25
2 Claim	2 under K 50
3 Negotiation	3 under K 100
4 Litigation	4 under K 500
5 Trial	5 under M 1
6 Settlement	6 under M 5
7 Arbitration	7 under M 10
8 Verdict	8 under M 50
9 Special ____	9 over M 50

Field yz: CENTRAL FACTOR (QF __ __)

1st digit: Activity

1 Planning, survey
2 Design, drawing
3 Specification
4 Fabrication, supply
5 Transport
6 Construction
7 Testing, inspection
8 Repair, rehabilitation
9 Use, occupation

2nd digit: Conclusion

1 Conditions contrary to assumptions
2 Inadequate design, detail for expected use
3 Mismanagement, rush
4 Clerical, copying error
5 Communication error
6 Negligent practice, violation of rules
7 Criminal economy
8 Intentional conduct
9 Defective materials, components

DATA COLLECTION AND INFORMATION DISSEMINATION: CURRENT EFFORTS AND CHALLENGES

By Donald W. Vannoy,[1] M. ASCE and Glenn R. Bell,[2] M. ASCE

ABSTRACT:

The need for the civil engineering community to collect and disseminate information on failures has been discussed widely. It appears that in recent years our profession's reluctance to deal actively with failures has passed, and tremendous interest in the concept of learning from failures has developed.

In pursuing the goal of a comprehensive system of failure information collection and dissemination we have faced several questions and problems: From whom will the information be collected? To whom must it be disseminated? What forms are required? How do we ensure that the disseminated findings are accurate and impartial? Can we obtain information on failures involved in litigation? How can we fund a comprehensive data collection and dissemination effort? Many of these questions can be answered by examining prior efforts.

Our assessment of our failure information needs is presented; present efforts at information collection and dissemination are reviewed; the challenges we face in developing a comprehensive system designed to allow us to learn from failures are discussed. The paper concludes with the authors' recommendations for ongoing efforts.

I. INTRODUCTION

Everyone agrees that the performance of our constructed facilities could be improved greatly if we had a systematic means to learn from failures. Not only the design/construction industry but also the general public would benefit. No one who calls himself a professional could oppose such a concept. Learning from failures, of course, means collecting and disseminating information on failures. What has been far less clear and debated widely is whether this should or even can be done. The issues are obvious: structures fail because people fail, and there is a natural reluctance to speak about the mistakes of others. Even if this reluctance is overcome in the interest of the greater good, a host of other problems arise. If inaccurate information is carelessly disseminated, innocent parties will be harmed.

---

[1]Associate Professor, Department of Civil Engineering, and Director, Architecture and Engineering Performance Information Center, University of Maryland, College Park, Maryland

[2]Associate, Simpson, Gumpertz & Heger Inc., Arlington, Massachusetts

In spite of the obstacles, information on failures has been disseminated. A very good summary of failure information disseminated from around 1875 to 1968 is given in Ref. (1). A summary of more recent efforts is given in Ref. (2). To the authors' knowledge, this information has been disseminated professionally and without harm to innocent parties. Nevertheless, no systematic method of information dissemination has existed. We have long recognized that as a profession we have been remiss in our obligation to learn from failures (3).

The verdict in the debate to disseminate or not to disseminate is in. The decision is in favor of dissemination, but it is qualified. Our reluctance to speak about failures appears to have passed. In the last five years, a great deal of interest has developed: the recent literature is filled with articles on the concept of learning from failures; several workshops and conferences have been held; many professional organizations are involved.

Two milestones in information collection and dissemination are the opening in 1982 of the Architecture and Engineering Performance Information Center (AEPIC) and the formation, also in 1982, of the ASCE Committee on Forensic Engineering. Based at the University of Maryland, AEPIC already is establishing satellite repositories in Great Britain and Canada. The ASCE Committee on Forensic Engineering has become the ASCE Technical Council on Forensic Engineering (ASCE-TCFE) and has formed several other committees. Among these are the Committee on Dissemination of Failure Information (ASCE-CDFI) and the Committee on Publications for the Journal of Performance of Constructed Facilities.

In pursuing the goal of a comprehensive system of failure information collection and dissemination we have faced several questions and problems. Many of these were anticipated; some were not:

- What forms of failure information are required?
- What are the sources of this information?
- To whom must this information be disseminated?
- How do we ensure that findings are accurate and impartial, so that innocent parties are not harmed?
- Can we disseminate information on failures involved in litigation?
- Who will fund the collection and dissemination effort?
- How do we coordinate the efforts of the many organizations interested in information dissemination?

This paper examines these questions. Our assessment of failure information needs is presented; past and present efforts at information collection and dissemination are reviewed; the challenges we are facing in developing a comprehensive system designed to allow us to learn from failures are discussed. The paper concludes with the authors' recommendations for ongoing efforts.

2.0 FAILURE INFORMATION NEEDS

2.1 Disciplines Requiring Failure Information

If the objective of disseminating failure information is to effect improved performance, then the answer to the question of who requires failure information is clear: it is those who can bring about positive changes in the design and construction process based on the lessons we learn from failures. Our tendency has been to focus largely on the Architect/Engineer (A/E) (4). This tendency is natural; most failures are perceived to be caused by errors in design, construction, or both. Design errors are thought to be the sole fault of the A/E, and construction errors are due, at least in part, to the A/E's failure to insure adequate quality control and quality assurance (QA/QC) over the construction. This mindset carries over from the traditional concept of the A/E as a master builder (4).

While we agree that A/E's, working within professional societies, should be leaders in initiating reforms and should regain greater control over the design/construction process (5), they cannot do it alone. The process by which we design and construct civil engineering facilities has become extremely complex. Others from within and outside of the design/construction industry must be involved, and, therefore, they also must have knowledge of past performance of engineered construction.

To appreciate the complexity of the information dissemination problem, one has only to consider some of the proposals under discussion for effecting changes in design and construction practice for improved performance (2, 6, 7, and 8). For example, questions of construction quality involve not only A/E's, but contractors, subcontractors, fabricators, materials suppliers, and producers of building components. Attorneys and insurance carriers must be involved to address questions of liability. Building officials, testing agencies, professional organizations, owners, and construction managers, need to be involved.

We need to bring the concept of failure into the classroom (9). Researchers and those who fund research in civil engineering need recognize the problems we face in performance of constructed facilities. The media, general public, and government officials who influence the construction industry (9 and 10) need information on failures.

Table 1, provides a comprehensive list of those disciplines requiring failure information.

TABLE 1

DISCIPLINES REQUIRING FAILURE INFORMATION

Attorneys	Materials Suppliers
Building Code and Specification Writers	Media
Building Officials	Owners of Constructed Facilities
Construction Managers	Policy Committees
Designers (Architects and Engineers)	Producers of Construction Components
Educators and Students	Professional Organizations
Fabricators	Researchers (University and Industry
Field Inspectors	Sources of Research Funds (Gov. & Industry
Forensic Architects and Engineers	Technical Committees
General Contractors and Subcontractors	Testing Agencies
General Public	
Government Officials	
Insurance Carriers	

2.2 Required Forms of Failure Information

To understand the required forms of failure information it is useful to have some appreciation for the types and causes of failures.

Failures can be catastrophic or subcatastrophic. While much attention has been focused on catastrophic failures, we have much to learn by studying the causes of subcatastrophic failures. We must use a very broad definition of failure. Neal FitzSimons uses three classes of failure as follows (12):

- A safety failure involves injury or death, or placing people in jeopardy.
- A functional failure involves the compromise of expected usage of a facility.
- An ancillary failure does not directly violate safety or compromise function, but it perversely affects schedules, costs or intended use.

Failures may occur in the preservice, service, or post-service phase of a facility.

The causes of failure can be classified as either Technical or Procedural (2 and 12):

- The Technical cause is the actual physical proximate cause of the failure.
- The Procedural cause is the human error, communication problem, or other shortcoming in the design/construction process that allowed the physical failure to occur.

It appears that most failures are caused by procedural errors rather than a lack of technical information (10).

With the increasing complexity of the design/construction process, the current legal environment, and fierce competition among designers and contractors, we find that communication problems, lack of definition of responsibility, and lack of continuity between parties are frequently responsible for failures (13). Thus, failure information that addresses interdisciplinary issues, and dissemination of that information to a broad base of disciplines is important.

There are as many required forms of failure information as there are disciplines that need it. Individual failure case studies are useful for educators and students, forensic architects and engineers, and researchers. Insurance carriers, professional organizations, government officials, and the general public need generic or trend-type analyses of issues in failures.

In 1984 a working group of the ASCE Committee on Forensic Engineering performed an extensive study of the role ASCE should play in the dissemination of information on failures of constructed facilities (14). The working group identified the following forms of failure information as the most needed:

- a journal of case histories of construction failures
- a regular column in a professional magazine (such as Civil Engineering)

- sessions at conventions
- individual case study publications
- seminars
- workshops
- short courses
- specialty conferences
- use of existing ASCE journals
- newsletters
- university and professional education strategies

As discussed in Section 3, examples of all of these forms of failure information already exist or are in the planning stage.

2.3 Sources of Failure Information

Whether the disseminated product of a failure study is an individual case study or a more general trend-type study, the product must be based on accurate, impartial information. Currently, the source of most failure information is individual consulting firms that investigate failures. Usually, investigations are performed directly for, or become involved with, some type of legal proceeding. This creates two problems:

- the accuracy or impartiality of findings can be questioned when the investigation is performed for an interested party
- the legal system sometimes impedes general release of such information

These problems have prompted proposals by some for an alternate means of investigating failures by an independent, impartial organization. Most notably, Gerald A. Leonards proposed a National Center to Investigate Failures (NCIF) in 1980 (15). The concept of a NCIF has been widely debated; time and space do not permit discussion of this here. The U.S. Congress has considered for some time the idea of using the National Bureau of Standards, Center for Building Technology (NBS/CBT) to conduct independent, impartial investigations of major structural failures (10 and 11). The NBS/CBT has conducted several such investigations of building catastrophies (16 through 20). In the foreseeable future, it appears that there will not be a NCIF, and the NBS will not be substantially more involved with investigation of failures than it is today. Our sources of failure information are not likely to change.

The problems associated with the current sources of failure information are not insurmountable, however. This is discussed in Section 4. In fact, as shown in Section 3, examples of nearly all of the required forms of failure information already exist in the literature. If we were only to benefit from the failure information already available, substantial progress would be made toward improved performance.

3.0 PRESENT COLLECTION AND DISSEMINATION EFFORTS

3.1 Architecture and Engineering Performance Information Center

The most extensive and comprehensive effort at collecting and disseminating failure information is the Architecture and Engineering Performance Information Center (AEPIC). AEPIC was founded at the University of Maryland in July 1982. The Center is a joint endeavor of the School of Architecture and the College of

Engineering at the University of Maryland, and was given its initial support by a two-year National Science Foundation grant. That grant, with additional support from the University of Maryland, Victor O. Schinnerer & Co., Sperry/Univac Corp., and others has now made it possible for the Center to enter the operational phase of its development.

AEPIC's data base covers performance information about buildings and civil structures, and includes all aspects of problems, such as the building envelope; structural, mechanical, and electrical systems; moisture barriers; economic and environmental concerns; as well as thermal, acoustical, visual, and behavioral dysfunctions. The Center's performance data relates to materials, elements, systems, processes and procedures. Factual information about cases currently under litigation is included in the files.

The data are stored in either computer data files or libraries, and include:

- **Computer "Performance Incident" or "Case" Files:** Professional and "informed reporter" reports on actual performance problems or malfunctions. Victor O. Schinnerer Co. has donated 40,000 claim reports to this file.

- **Computer "Citation" Files:** References to published information about performance problems that have appeared in journals, trade press magazines, newspapers, and agency investigation reports. This file currently includes Engineering News Record articles for the last 20 years.

- **Dossier Library:** Documentation of performance data about the incidents and related information in the " Case" files.

- **Visual Materials Library:** Photographs, slides, and other visual materials related to the "Case" files.

- **Reference Library:** Current and historical codes, standards, and other technical references.

Some of the significant features of the AEPIC information, in connection with earlier discussion on required forms of failure information, are as follows:

- Information in the data base is free of sensitive personal information to protect the privacy of individuals and firms.

- Opinions on procedural causes of failure and culpability are collected.

- The information in each Case File is raw; i.e., it has not been distilled, and is dependent on the competency and objectivity of the reporter, and on any possible limitations on the investigation reported.

The University of Maryland serves as the International Center and the National Repository of AEPIC, and will be augmented by international repositories that will broaden the base of AEPIC users and contributors. The most advanced international repositories are in Canada and the United Kingdom.

An Advisory Board of nine professional leaders provides advice and guidance on AEPIC policies, programs, and technical operations. The Advisory Board members were selected for expertise in architecture, engineering, engineering testing, geotechnical analysis, insurance, law, contracting, and research.

In addition, an Advisory Council and nine Advisory Committees have been formed to provide liaison between AEPIC and membership organizations, technical and trade associations, councils and institutes, and major agency media, research, educational, legal, and technical user networks. Approximately 150 members of the Council and Committees are assisting in dissemination of information about AEPIC, encouraging their members to contribute data to AEPIC, and reviewing AEPIC functions

AEPIC is a non-profit center. An annual membership fee, in addition to a search fee, is charged to maintain the data base. A query is conducted by "keyword" and the results of the search are sent to the inquirer. Copies of the citation or case forms, which match the "keywords" provided, are sent to the inquirer. Annual memberships are available for individuals, firms, organizations, and corporations.

Readers interested in further background on AEPIC may see Refs. (21 and 22).

3.2 U. S. Committees and Organizations

Although AEPIC and some of the other groups involved in learning from failures are relatively new, professional organizations have been disseminating information on failures for decades. A listing of the more prominent of these is given below; further information on many of them is given in Ref. (2).

- ACI Committee 437, Strength Evaluation of Existing Concrete Structures
- American Consulting Engineers Council
- ASCE Engineering Performance Information Committee (EPIC)
- ASCE Technical Council Forensic Engineering (TCFE)
  TCFE presides over four technical committees:
    - Committee on Dissemination of Failure Information (CDFI)
    - Committee on Practices to Reduce Failures
    - Task Committee on Guidelines for Failure Investigations
    - Committee on Publications for the Journal of Performance of Constructed Facilities
- ASCE Pipeline Division-Committee on Forensic Engineering
- ASCE Technical Council on Research (TCOR)
- Association of Soil and Foundation Engineers (ASFE)
- The Engineering Foundation
- Performance of Structures Research Council (PSRC) of TCOR
- National Academy of Forensic Engineers (NAFE)
- National Academy of Science-National Research Council (NAS/NRC), U. S. National Committee
- National Society of Professional Engineers (NSPE)
- Raymond DiPasquale & Associates, Information Systems Division
- United States Committee on Large Dams (USCOLD)

3.3 Foreign Efforts

The ASCE-TCFE and the ASCE-EPIC correspond with foreign organizations involved in disseminating failure information, and, as mentioned above, AEPIC is establishing foreign repositories.

In 1985, The Construction Performance Centre was established in the United Kingdom. It will consist of a number of Construction Performance Units coordinated by an Advisory Board. The Units will collect information for a computer data base and assist users of the service. Other information sources in the United Kingdom and abroad will be linked with the Centre. Presently, Units have been formed at the University of Manchester Institute of Science and Technology and at the University of Strathclyde in Glasgow. The United Kingdom Building Committee of the Science and Engineering Research Council has awarded a grant to establish the necessary computer hardware and software for the data base.

Other foreign efforts in disseminating failure information have been primarily limited to seminars, lectures, and individual papers. A typical example is the seminar "Lessons from Failures of Structures" organized by the Maharashtra, India Chapter of American Concrete Institute on December 17-19, 1982 in Bombay, India. The proceedings have been donated to the AEPIC files.

A systematic international reporting network needs to be established and maintained to collect and disseminate failure information and to access failure information that may be available in other foreign countries.

3.4 Examples of Information Dissemination Efforts

While many consider that little has been done to disseminate information on failures, and that the problems involved prohibit doing so, the available literature contains examples of nearly every form of failure information mentioned in Section 2.2. A review of some of these examples is useful for planning failure dissemination efforts. Further examples and a more in-depth discussion are included in Ref. (2).

Books

While there are no textbooks on construction failure, there is a handful volumes giving brief case histories. These include:

- Construction Failure by Jacob Feld (1): probably the best known of such volumes
- Structural and Foundation Failures by Barry B. LePatner and Sidney Johnson (23)
- Concrete Problems, Causes and Cures by John C. Ropke (24)
- Construction Disasters: Design Failures, Causes, and Prevention by Steven Ross (25): based on several years' articles from Engineering News Record

To Engineer is Human by Henry Petroski (26) is a more reflective volume dealing with the procedural aspects of failure.

Educational Media

Courses on forensic engineering have been taught at a handful of universities (e.g., University of Delaware (21)). However, at this time there exists no textbook or generally available organized curricular materials.

Conferences, Conventions, and Short Courses

The American Concrete Institute widely disseminated failure case studies through its seminars entitled "Lessons from Failures of Concrete Buildings" (28). Recent conferences dealing with improving practice based on the lessons learned from failures include The Engineering Foundation Conference held in Santa Barbara in November 1983 and the NSF/ASCE conference "Reducing Failures of Engineered Facilities" held in Clearwater Beach, Florida in January 1985.

Monograph Series and Case History Publications

Excellent examples of collections of case histories are those by the Association of Soil and Foundation Engineers and by the United States Committee on Large Dams (29 and 30).

Dozens of individual case histories have been published in the professional journals. A few examples are given in Refs (29 through 39). The NBS's reports of structural failures, mentioned in Section 2.3 above, are available to the general public.

Journal of Forensic Engineering

A professional journal devoted solely to performance problems of constructed facilities has been under discussion at AEPIC, ASCE, and other professional organizations for some time. A key characteristic of such a journal is that it be interdisciplinary in content and distribution. The journal should include not only case history articles, such as given in (31 through 39), but also discussion of generic issues or trend analyses in failures such as given in (40 through 43).

Plans for such an interdisciplinary journal, to be published by the ASCE, are well advanced, and ASCE-TCFE has formed a publications committee for such a purpose. NSPE and AEPIC have indicated interest in participating; it is hoped that additional organizations will join the effort.

4.0 PROBLEMS IN COLLECTING AND DISSEMINATING FAILURE INFORMATION

4.1 Coordination of the Data Collection and Information Dissemination Effort

The ideal data collection and information dissemination machine would collect failure information from the sources given in Section 2.2, extract, process, organize, and produce this information in the needed forms described in Section 2.3, and disseminate it to those end users mentioned in Section 2.1. Our current system falls far short of this ideal.

A wealth of good accurate failure information exists and is available for dissemination, and many organizations, such as those mentioned in Section 3, are anxious to disseminate it. A lack of organization and coordination has prevented this information from being widely disseminated and put to good use, however. We must recognize that each organization interested in the causes of failures approaches the subject with different goals. Some address only procedural problems; some address very technical information of limited breadth. All may contribute to the dissemination process, but their efforts must be well directed. At present the authors foresee the following needs to ensure that this occurs:

- Participating organizations must communicate, cooperate, and share information more readily.
- Participating organizations must better define their respective responsibilities and areas of interest.
- A dominant, coordinating organization should be identified.

4.2 Ensuring Accuracy and Impartiality of Findings

Construction failures are devastating occurrences. At best, they are expensive to correct. At worst, people can be injured or killed. The inevitable lawsuits that result from failures are long, expensive events, which can be grueling for those involved. Dissemination of information on failures cannot be taken lightly. If inaccurate information is carelessly released, innocent parties can be harmed and slander suits can result.

Some have advocated not discussing the procedural causes of failures in disseminated information, in an attempt to avoid assigning responsibility for failures. We believe that because procedural problems are the major contributor to failures, they must be discussed. Names and personal information of involved parties can be omitted from discussions, but this will not ensure confidentiality. The parties involved in major failures are generally known through publicity, or such information can be obtained easily by other means for lesser failures.

Complete accuracy of findings cannot be guaranteed, however. Even the most competent and objective of investigators frequently disagree on the causes of failures. Disseminators of failure information cannot be the omnipotent judges and juries. Room for disagreement must exist. The well-known case of the Hyatt Regency walkway collapse is one in which there is substantial disagreement in the industry over a procedural problem: responsibility for structural steel connection design. In spite of this disagreement, however, information on this failure has been widely and professionally published, and as a result, professional committees are working to resolve a problem in our industry.

What is required in any published article on construction failure is the highest standard of professionalism and objectivity. To the extent possible, accuracy of findings must be ensured. We believe the essential components of editorial policy are as follows:

- Information must be based on a well-founded investigation.
- The reporting investigators must be qualified.
- The reported results must be based on factual information - hearsay and speculation must be eliminated.
- All articles should be subjected to a rigorous review process.
- Where there is disagreement over findings or an opposing point of view, all issues should be discussed.
- The extent to which procedural errors may be discussed should vary depending on the source, quality, and type of available information.

Data in the AEPIC files are raw; they are entirely dependent on the objectivity and quality of the investigation on which they were based. The AEPIC user must understand this, and use the information accordingly.

4.3 Supporting the Collection and Dissemination Effort

A comprehensive data collection and dissemination effort is a costly undertaking, but the potential rewards are great. A significant problem in obtaining resources for these efforts is that the rewards are long-term. Even after failure information is disseminated, it takes years to effect improved practice. There is no obvious short-term gain for a contributor.

Progress will continue to be made through volunteer efforts, mostly professionals working through professional societies. But this is not enough. Presently we have an established information repository in AEPIC. Industry-wide support of AEPIC and contributions of failure information to it are essential if AEPIC is going to flourish, however.

Additional financial support must be obtained if AEPIC is to survive until the data base is large enough to make it self-supporting, if it ever can be. Unfortunately, there are few libraries that are totally self-supporting. Support should go beyond the architectural and engineering community to include all those who stand to gain from improved performance, such as building owners, insurers, and the general public.

In addition to financial support, the contribution of data to AEPIC is the key to the network of disseminating failure information. Unless the practitioner takes time to fill out the reporting forms and return the reporting forms with the appropriate attached reports, drawings, and other materials, the complete system will never be established.

The question for the reader is, "Have you taken the time to report one case to AEPIC this year?"

4.4 Constraint of the Legal System on Releasing Failure Information

Many have expressed the belief that our legal system prohibits general release of information on failures. Our experience has shown that this is largely a myth. It is true that while a case is under litigation, a party who is investigating a failure on behalf of one of the litigants is bound by confidentiality. There is good reason for this. However, once the litigation has settled, information usually may be released. One of the authors (Bell) has been generally successful in obtaining permission to release failure information after the case is closed. To our knowledge, there has been only one well published case where engineering experts were bound to confidentiality after the lawsuit was settled. Typically, it may take several years for a case to be resolved, but this doesn't make the information useless. The most important lessons we have to learn from failures today are based on mistakes that have been made repeatedly over many years. If we were only to disseminate the wealth of information that is available for release today, a great deal of progress could be made.

5.0 SUMMARY AND RECOMMENDATIONS

We have much to learn from failures of engineered construction, and the recent interest that has developed in disseminating failure information promises that this will be done. While there are some significant challenges in the endeavor, examination of prior efforts shows that it can be done.

The following are the authors' recommendations for future efforts:

(1) We should broadly disseminate information on failures of constructed facilities to people who can effect improved performance. These people include those within and outside of the design/construction industry.

(2) The forms of failure information we disseminate should vary widely from case-specific studies to general trend-type analyses of failures. More emphasis is required on the following:

- disseminating information on minor failures, near failures, serviceability problems, and function impairment
- identifying the procedural causes of failures
- disseminating information that is interdisciplinary

(3) We should work with the current sources of failure information. Much information on cases involved in litigation may be released eventually.

(4) We must support and encourage the further development of AEPIC. The two most critical issues in AEPIC's growth are the following:

- broad-based financial support
- contributions of failure information by forensic practitioners

(5) We should continue the development of the proposed Journal of Performance of Constructed Facilities, which should include case studies as well as generic failure information. The Journal must be based on strict editorial policy that will ensure, to the extent possible, accuracy, confidentiality, and impartiality to all involved parties.

(6) We must effect better liaison and coordination of various organizations, foreign and domestic, that are concerned with disseminating failure information.

(7) We should work toward a long-term goal of developing a communications network that will ensure that all parties who can benefit from improved performance receive it in needed form.

APPENDIX - REFERENCES

(1) Feld, Jacob, Construction Failure, John Wiley and Sons, New York, 1968.

(2) Bell, Glenn R., "Failure Information Needs in Civil Engineering," Reducing Failures of Engineered Facilities, Proceedings of a Workshop Sponsored by the National Science Foundation and the American Society of Civil Engineers, Clearwater Beach, Florida, January 7-9, 1985, American Society of Civil Engineers, New York, 1985.

(3) American Society of Civil Engineers, Research Council on Performance of Structures, Structural Failures: Modes, Causes, Responsibilities, American Society of Civil Engineers, New York, 1973.

(4) Dib, Albert, "Liabilities of the Contributors to the Construction Process," 1985 Legal Handbook for Architects, Engineers, and Contractors, Albert Dib and James K. Grant, Ed., Clark Boardman Co., Ltd., New York, 1985.

(5) Pfrang, Edward O., "Structural Failures and Professional Liabilities," Civil Engineering, American Society of Civil Engineers, Vol. 54, No. 12, New York, December, 1984.

(6) Haines, Daniel W., "Forensic Engineering: What Role for ASCE?," Civil Engineering, American Society of Civil Engineers, Vol. 53, No. 7, New York, July, 1983.

(7) Godfrey, K. A., Jr., "Building Failures -- Construction Related Problems and Solutions," Civil Engineering, American Society of Civil Engineers, Vol. 54, No. 5, New York, May, 1984.

(8) Hohns, H. Murray, "Procedural Changes in the Design and Construction Process to Reduce Failures," Reducing Failures of Engineered Facilities, Proceedings of a Workshop Sponsored by the National Science Foundation and the American Society of Civil Engineers, Clearwater Beach, Florida, January 7-9, 1985, American Society of Civil Engineers, New York, 1985.

(9) Chesson, Eugene, Jr., "Failure Investigation in Undergraduate Education," Reducing Failures of Engineered Facilities, Proceedings of a Workshop Sponsored by the National Science Foundation and the American Society of Civil Engineers, Clearwater Beach, Florida, January 7-9, 1985, American Society of Civil Engineers, New York, 1985.

(10) Gatje, Scott P., "The Role of the Federal Government in the Investigation of Structural Failures," WISE Internship Report to the American Society of Civil Engineers, August 3, 1984.

(11) U. S. House of Representatives, Committee on Science and Technology, Subcommittee on Investigation and Oversight, "Structural Failures," Hearing Before the Subcommittee on Investigation and Oversight, April 25, 1984, U. S. Government Printing Office, Washington, 1984.

(12) FitzSimons, Neal "Notes on Statistics of Failures of Constructed Works," Reducing Failures of Engineered Facilities, Proceedings of a Workshop Sponsored by the National Science Foundation and the American Society of Civil Engineers, Clearwater Beach, Florida, January 7-9, 1985, American Society of Civil Engineers, New York, 1985.

(13) Thornton, Charles H., "Failure Statistics Categorized by Cause and Generic Type," Reducing Failures of Engineered Facilities, Proceedings of a Workshop Sponsored by the National Science Foundation and the American Society of Civil Engineers, Clearwater Beach, Florida, January 7-9, 1985, American Society of Civil Engineers, New York, 1985.

(14) American Society of Civil Engineers Committee on Forensic Engineering, "Report of the Working Group on Information Dissemination --Committee on Forensic Engineering -- ASCE," March, 1984.

(15) Leonards, Gerald A., "Investigation of Failures," Journal of the Geotechnical Engineering Division, Proceedings of the American Society of Civil Engineers, Vol. 108, No. GT2, New York, February, 1982.

(16) U. S. Department of Commerce, National Bureau of Standards, Investigation of Construction Failure of Harbour Cay Condominium in Cocoa Beach, Florida, NBS Building Science Series 145, U. S. Government Printing Office, Washington, D.C., August, 1982.

(17) U. S. Department of Commerce, National Bureau of Standards, Investigation of the Kansas City Hyatt Regency Walkways Collapse, NBS Building Science Series 143, U. S. Government Printing Office, Washington D.C., May, 1982.

(18) U. S. Department of Commerce, National Bureau of Standards, Investigation of Skyline Plaza Collapse in Fairfax County, Virginia, NBS Building Science Series 94, U. S. Government Printing Office, Washington, D.C., February, 1977.

(19) U. S. Department of Commerce, National Bureau of Standards, Investigation of Construction Failure of Reinforced Concrete Cooling Tower at Willow Island, West Virginia, NBS Building Science Series 148, U. S. Government Printing Office, Washington, D.C., September, 1982.

(20) U. S. Department of Commerce, National Bureau of Standards, Investigation of Construction Failure of the Riley Road Interchange Ramp, East Chicago, Indiana, NBSIR 82-2593, U. S. Government Printing Office, Washington, D.C., October, 1982.

(21) FitzSimons, Neal and Vannoy, Donald, "Establishing Patterns of Building Failures," Civil Engineering, American Society of Civil Engineers, Vol. 54, No. 1, New York, January, 1984.

(22) Vannoy, Donald W., "Report to Executive Director, American Society of Civil Engineers, from Architecture and Engineering Performance Information Center," New York, November 12, 1984.

(23) LePatner, Barry B., and Johnson, Sidney M., Structural and Foundation Failures: A Casebook for Architects, Engineers, and Lawyers, McGraw-Hill, New York, 1982.

(24) Ropke, John C., Concrete Problems: Causes and Cures, McGraw-Hill, New York, 1982.

(25) Ross, Stephen S., Construction Disasters: Design Failures, Causes, and Prevention, McGraw-Hill, New York, 1984.

(26) Petroski, Henry, To Engineer is Human, St. Martin's Press, New York, 1985.

(27) Chesson, Eugene, "How Not to Do It!," Proceedings, National Conference on Engineering Case Studies, March 28-30, 1979.

(28) American Concrete Institute, "Seminar Course Manual: Lessons from Failures of Concrete Buildings," American Concrete Institute, undated.

(29) United States Committee on Large Dams, Lessons Learned from Dam Incidents --USA, 1975.

(30) United States Committee on Large Dams, Lessons Learned from Dam Incidents, 1979.

(31) Smith, Erling A., and Epstein, Howard I., "Hartford Coliseum Roof Collapse: Structural Collapse Sequence and Lessons Learned," Civil Engineering, American Society of Civil Engineers, Vol. 50, No. 4, New York, April, 1980.

(32) Schousboe, Ingvar, "Bailey's Crossroads Collapse Reviewed," Paper 12186, Journal of the Construction Division, Proceedings of the American Society of Civil Engineers, Vol. 102, No. CO2, New York, June, 1976.

(33) Fairweather, Virginia, "Bailey's Crossroads: A/E Liability Test," Civil Engineering, American Society of Civil Engineers, Vol. 45, No. 11, New York, November, 1975.

(34) Pfrang, Edward O., and Marshall, Richard, "Collapse of the Kansas City Hyatt Regency Walkways," Civil Engineering, American Society of Civil Engineers, Vol. 52, No. 7, New York, July, 1982.

(35) Hauck, George F.W., "Hyatt Regency Walkway Collapse: Design Alternatives," Journal of Structural Engineering, Proceedings of the American Society of Civil Engineers, Vol. 109, No. 5, New York, May, 1983.

(36) Hahn, Oscar M., "Stability Problems of Wood Truss Bridge," Paper 7095, Journal of the Structural Division, Proceedings of the American Society of Civil Engineers, Vol. 96, No. ST2, New York, February, 1970.

(37) Nordlund, Raymond L., and Deere, Don U., "Collapse of Fargo Grain Elevator," Paper 7172, Journal of the Soil Mechanics and Foundations Division, Proceedings of the American Society of Civil Engineers, Vol. 96, New York, March, 1970.

(38) Lew, H. S., " West Virginia Cooling Tower Collapse Caused by Premature Form Removal,"Civil Engineering, American Society of Civil Engineers, Vol. 50, No. 2, New York, February, 1980.

(39) Bell, Roy A., and Iwakiri, Jun, "Settlement Comparison Used in Tank-Failure Study," Paper 15219, Journal of the Geotechnical Engineering Division, Proceedings of the American Society of Civil Engineers, Vol. 106, No. GT2, New York, February, 1980.

(40) Vannoy, Donald W., "20/20 Hindsight -- Overview of Failures," Proceedings of the Conference "Construction Failures: Legal and Engineering Perspectives," American Bar Association, October, 1983.

(41) Thornton, Charles H., "Lessons Learned From Recent Long Span Roof Failures," Notes from ACI Seminar on Lessons from Failures of Concrete Buildings, Boston, April 6, 1982.

(42) Stockbridge, Jerry G., "Cladding Failures -- Lack of a Professional Interface," Paper 15085, Journal of the Technical Councils, Proceedings of the American Society of Civil Engineers, Vol. 105, No. TC2, New York, December, 1979.

(43) Reese, Raymond C., "Structural Failures from the Human Side," Civil Engineering, American Society of Civil Engineers, Vol. 43, No. 1, New York, January, 1973.

CONSTRUCTION INSURANCE: AN ALTERNATIVE UNIFIED RISK INSURANCE

C. ROY VINCE, CPCU, ENVIRONMENTAL LAW INSTITUTE*

INTRODUCTION

The construction industry is an industry that is growing used to conflict as an almost inherent part of its nature. Cooperation and efforts to meet the construction goal are negated by concern over construction liabilities. In fact, the preparation for dispute and ultimate assessment of liability at the outset of the project is more the rule than the exception.

The insurance industry has traditionally delivered its insurance products in two forms, property protections and liability protections. Property protection provides payment in case an insured event happens and damage occurs for which protection is arranged. Liability protections pay on behalf of an insured for damages he causes others for which he is legally liable.

The liability protections blend perfectly with the construction industry's inherent confrontation and risk taking nature. Liability insurance requires the proof of legal responsibility. Conflict is a necessity. Very little protection, however, is offered by these policies The passion to prove fault becomes an economic necessity. Tremendous resources are allocated to the process of establishing fault.

The aftermath of a construction failure or deficiency will be confrontation to determine fault. At first glance, one might define fault finding as a search for the cause of failure. More than this is at stake. The search is for the legally responsible party, the party who will pay for the damages to property, persons and to the organizations associated with the construction process who have been damaged as a result of the failure. The cause of the failure often becomes obscure in fulfilling the obsession to establish liability.

*Vice President, Professional Liability Brokers & Consultants, Inc., 1011 E. Touhy Avenue, Suite 321, Des Plaines, IL 60018.

OVERVIEW

The primary property type protection on the job is the builder's risk form. Usually this coverage excludes loss resulting from construction failure. The liability policies available to contractors do not afford coverage for work in progress. The coverage that is afforded is based on the proof of legal liability and extends to persons injured who suffer bodily injury or property damage. The property damage liability would typically apply in the event of a construction failure, to property of the public, not the property of those engaged in the project.

The injured worker would have the right to worker's compensation for his injury. In most jurisdictions he would be barred from any other recourse against his employer but would have a right of action against all other participants in the construction process not protected by the worker's compensation statute. Outsiders to the construction process who suffer bodily injury or property damage can bring suit, selectively, against the principals involved in construction or sue everyone. Typically, the constructor would make denials and engage in a series of cross claims against the other construction participants.

The serious construction deficiency would typically trigger massive litigation attendant to the process of fixing legal liability. This process creates a massive drain on resources. The injured and the estates of the deceased will typically file suit against each party who may conceivably have had some degree of responsibility for the failure. Each of those defendants will engage in the legal ritual of denying their fault and raise questions concerning the fault of their co-defendants. Each party to the construction process will, in some way, be damaged and additional suits will follow demanding payment for physical damages as well as economic loss. Counter-claims and cross claims typically will follow, complicating the legal labyrinth.

The cost of this approach is not easily calculated but is readily imaginable. Insurance bears only a small portion of the costs; the lion's share falls to the construction participant. One pays with his time and energies to deal with these uncertainties. Litigation is expensive. Contribution to settlement is often made to avoid being adversely affected by the whims of the court. These costs are multiplied by the number of the participants caught in the litigation maze.

There is no ready authority to measure this drain on the construction industry. An assessment of the scope of the problem and knowledge of the elements of cost leads one to conclude a substantial portion of the construction dollar goes to determining liability.

One observation lends insight to the scope of the cost. The design professional is typically insured for professional liability which will pay his legal costs in defending himself from such claims. About one-third of the premium dollar used to purchase professional liability insurance pays for the cost of establishing fault. That price tag can be estimated at over $200 million dollars per year. Applying multiples of that number to estimate the costs to the entire industry affected leads one to conclude that in affixing liability, the construction industry pays billions of dollars annually.

Can we afford the cost? We certainly can. The costs have been with us; we've managed and we can continue to do so. Can we avoid the cost? Certainly, but there will be required an enormous effort, and the coordinated cooperation of diverse groups. A reallocation of construction risk is necessary to substantially address the problem. The insurance community should be able to accept coverage for additional insurable events. The construction participants need to contract to fairly assume uninsurable risk. Unfortunately, such a solution is not immediately at hand. The current thinking of both the construction and insurance community would have to undergo radical change.

It is most difficult to cause change when divergent self-interests and tradition have to be abandoned. A new coalition of common interest would have to be constructed to set goals and plan the steps to cause change.

The solution goes beyond the problem of dealing with failure. It is a solution that addresses resolving the bodily injury, property damage and consequential loss arising from a deficiency in design or construction.

Some considerable work is underway. In London, in June of 1978, the Independent Federation of Consulting Engineers met with a group of international consulting engineers, contractors, owners, insurance underwriters, brokers and certain officials from international lending agencies to discuss this problem. One of the concepts discussed was a scheme under which the owner takes out all insurance protection covering both contractor and engineers. Therefore, it would not be required to prove whether a contractor or engineer is to blame when something adverse occurs. It is not known whether any subsequent work was done by this group in pursuing the concepts discussed there.

A November, 1983 Engineering Foundation Conference on Structural Failures concluded that there should be a form of unified all risk insurance to protect all members of a construction team. This writer developed such an insurance concept that was adopted by an Engineering Foundation conference in January, 1984. The American Consulting Engineers Council, at its fall conference in Little Rock, Arkansas, followed in the path of the prior conferences. At an American Society of Civil Engineers workshop held in Chicago in November, 1984 the workshop committee reiterated the conclusions of the prior conferences. The concept is being referred to as Unified All-Risk Insurance. The ICED Joint Committee on Professional Liability, at its July 25, 1985 meeting, studied _A Proposed Program of Construction Insurance_ authored by this writer. They approved the plan with minor modifications and are planning for its implementation.

In order to be effective, the concept will have to be translated into a market demand. The market demand would have to be founded on a conviction that it is of economic benefit to purchase insurance to insure all construction participants to avoid the cost of apportioning legal liability among the construction participants.

A market demand is an important incentive for the insurance industry to assume these risks.

INSURANCE PRINCIPLES

Insurance is defined as a device for reducing risk by combining a sufficient number of exposure units to make the individual loss collectively predictable. The criteria of insurability is that:

1. There is a large number of homogeneous exposure units involved.

2. The loss produced by the peril must be definite.

3. The occurrence of the loss in the individual case must be accidental and fortuitous.

4. The potential loss must be large enough to cause hardship.

5. The cost of the insurance must be economically feasible

6. The chance of loss must be calculable.

7. The peril must be unlikely to produce loss to a great many insured units at one time.

Two problems of the construction industry are capable of a solution.

1. The bodily injury to workers and others and property damage to non-construction participants from construction failure or construction accident.

2. The property damage and consequential loss to construction participants from failure.

Both problems can be resolved by designing insurance products tailored to meet construction industry needs. Insurance should be purchased by the owner to cover himself and all participants governed by the construction documents.

CONSTRUCTION LIABILITY INSURANCE

The first problem is easily solved by the use of traditional insurance protections. A Construction Liability policy could be written to cover all the construction participants for all but worker's compensation liability and vehicular liability. This policy would cover bodily injury to persons arising from the construction process. Property damage liability protection would be afforded to all construction members for their liability to non-members of the construction team. Historically, some wrap-up type insurance schemes have approached achieving this result. Although the liability of one or more members of the team would have to be established in the process of compensating for the injury or damage, the problem of dealing with the inter-team member disputes could be eliminated.

CONSTRUCTION PROPERTY INSURANCE

The problem of dealing with the claims of one team member for his damages against other team members is not capable of such easy solution. The solution of this problem could save the construction industry billions of dollars a year from the avoidance of assessing fault for construction failures. These losses can best be insured by taking a property insurance approach rather than a liability approach.

This concept contemplates a comprehensive policy that would, when certain events take place, compensate the damaged party without the necessity to prove legal liability. Substituted would be the principles of the first party coverage, i.e. an insured event takes place, and the payment is made in accordance with defined rules.

A purpose of the proposed Construction Property Insurance would be to avoid litigation among the parties which essentially arises from the failure of first party protections to cover events on the construction project which result in loss. The first party coverages generally exclude loss caused by, or resulting from, error, omission, or deficiency in design, specifications, workmanship or materials. This exclusion, however, does not apply when the following perils occur: loss by fire, lightning, windstorm, hail, explosion, riot, civil commotion, aircraft,vehicles, smoke or discharge from protection or building service equipment. Personal property forms in place usually exclude actual work upon, installation or testing of property covered, failure, breakdown or derangement of machines or machinery, error, omissions, or deficiency in design specifications, workmanship or materials, unless loss by fire or explosion not otherwise excluded ensues. Also excluded are delay, loss of market, interruption of business, or consequential loss of any nature. Also excluded is water which backs up through sewer or drains.

Liability policies would apply only if legal liability occurred and if the event met the definition of an occurrence which usually is defined as "An accident including continuous or repeated exposure to conditions which results in bodily injury or property damage, neither expected nor intended, from the standpoint of the insured". In order for coverage to exist in a liability policy, there must be some type of bodily injury or physical damage to property. The policies generally exclude loss of use of tangible property which has not been physically injured or destroyed resulting from a delay and/or lack of performance by, or on behalf, of the named insured of any contract or agreement, or the failure of the named insured's products, or work performed by or on behalf of the named insured to meet the level of performance, quality or fitness or durability warranted or represented by the named insured. Also excluded is property damage to the named insured's products arising out of such products or any part of such products. Also excluded is property damage to work performed by or on behalf of the named insured arising out of the work or any portion thereof, or out of materials, parts or equipment furnished in connection therewith. The professional liability policy of the design professional requires proof of legal liability which translates to a breach of the standard of care. This imposes a considerable burden of proof upon any party wishing to make this liability claim.

It is these characteristics of the builder's risk policy, the property protection, the general liability policies, and the professional liability policy which sets the stage for construction disputes between construction participants when the material deficiency or failure occurs.

INSURING AGREEMENTS

The proposed Construction Property Insurance, by insuring these exclusions under appropriate conditions would substantially eliminate the need for construction disputes. The format of the construction protection would be to insure all the parties who are held to the general conditions of the construction contract. The insuring agreement would insure each of these parties for the loss due to a physical defect caused by a product failure, a workmanship failure, or a design failure. The policy could be endorsed to extend the basic coverage for certain consequential losses, such as delay, loss of business, or interruption of business.

POLICY EXCLUSIONS

The exclusions would exclude coverage under this policy for coverages otherwise provided under the builder's risk cover, the property and liability and worker's compensation coverages which the parties to the construction process would normally have purchased. In addition, a series of exclusions would be necessary to deal with the moral hazard inherent in the construction process of knowingly providing defective products and materials, knowingly faulty installation, knowingly substandard workmanship, knowing violation of building code requirements, intentional misdesign, and intentional alteration of construction for financial benefit. Also excluded would be warranties and guarantees. An additional endorsement could be considered within confined parameters to endorse coverage to meet a certain performance criteria in those projects where one is applicable.

UNDERWRITING CRITERIA

The policy would have to be carefully underwritten. Since it would be underwritten on a project basis, and the coverage issued pursuant to the terms of the construction contract, the timing of the coverage would be important. Construction begins after the execution of the construction contracts. Prior to this time, construction documents have been prepared. It is proposed that appropriate modifications be made to the general conditions of the construction documents. The Unified Risk Insurance should be put in place in conjunction with the signing of the construction documents, and the commencing of work.

The construction insurance policy would be written with a

policy limit in the aggregate for all insureds, a sum, not to exceed the cost of the construction project. Normally, the limit of liability would be a fixed amount, arbitrarily determined or a fixed amount which would be a percentage of the total cost of construction. The carrier's liability, at the time the claim was made, would not exceed the value of the project at the time of the occurrence of the covered event. Each insured would have a deductible equivalent to some predetermined sum, at lease sufficient to impose a burden, but not so large as to pose an impossible hardship.

The underwriter, as part of the underwriting process, would review, carefully, the credentials of the construction participants and base the decision to enter into such a contract on past history, reputation and financial information of the participants. Of primary consideration would be past claim history and litigation history of the participants. Specifically, the underwriter of such coverage would need to review the contract to determine whether or not appropriate contractual provisions were entered into to foster success. The construction documents would be reviewed by the insurer's engineering department to assure that the construction documents were sufficiently complete for the purpose of the insuror attaching to the risk. During construction, inspection by the design professionals involved would be required. The owner would provide these to the insurer who would utilize them in the ongoing underwriting process.

The coverage would terminate at the time of completion of the construction phase. Subsequent coverage could be obtained to continue protection for the owner during occupancy.

CONTRACTS FOR CONSTRUCTION

The appropriate contracts between the parties, i.e. the owner/architect agreement, architect/engineer agreement, owner/contractor agreement, contractor/subcontractor agreement, need to be modified to create a receptive climate for the underwriter to issue Unified Risk Insurance and to provide the additional necessary elements to avoid the construction dispute. The contract documents that are used in the bidding process would have to incorporate the necessary changes.

Arbitration or a format such as mediation-arbitration should be an available remedy to resolve questions of specific performance under the contract. Insurance dispute resolution would be governed by the insurance contract. The dispute provision should reinforce the limitation of other legal remedies.

DISPUTES REMAINING

The aforementioned insurance coverages outlined and the changes to the contract documents would cause a shift in the entire dispute process. Only disputes that deal with the claims of parties not included in the construction contract process, such as workers, passersby, trespassers, and adjoining property owners would be concluded using the traditional process of litigation Disputes that would normally involve contract, tort and warranty litigation between the participants in the construction process could be eliminated by a partial shifting of the burden to an insuror who would insure, on a first party basis, events that would normally trigger litigation. The parties would assume a known risk within a manageable framework, the deductible. The parties would be limited to their recovery for damages to the insurance protections they are able and willing to secure. The contract disputes would be resolved by an alternate means of dispute resolution that would be confined to specific performance. Performance would be assured by the surety protections. Dispute resolution would be timely and efficient.

SUMMARY

The transfer of risk mechanism (insurance) can be applied to solve problems unique to the construction environment. The proposed program is worthy of support since all segments of the construction industry would benefit.

The rights of the users and public would be unaffected since, as non-parties to the construction contract, all of their legal rights and remedies would be preserved. The insurance contracts currently available to the construction participants would need minor modification to apply to such claims. Such a proposal would provide a reasonable framework with known assumed and allocated risks for the participants in the construction project and not diminish in any way the rights of the general public.

In the event of a failure, the parties could mutually set about determining the cause of the failure to expand their knowledge and avoid future failures, rather than attempt to protect their positions so as to avoid being assessed the liability. Avoided would be the waste of resources attendant to the process of assessing liability among construction participants.

The questions that remain: Is the construction industry desirous of supporting this program through premium payments to spread the risk? Is the insurance industry willing to offer such products tailored to the need and demand of the construction industry?

Major national construction organizations are now engaged in a considerable effort to obtain the answers to these questions.

QUALITY ASSURANCE/QUALITY CONTROL:
ALTERNATIVE APPROACHES TO REDUCE FAILURES

James M. Hinckley, Member ASCE*

ABSTRACT:

Efforts to achieve quality on construction projects usually center around code-compliant design, thorough specifications, and careful inspection. The alarming increase in both the frequency and severity of construction problems, abnormal structural behavior, and partial or total collapse indicates that these commonly employed measures are inadequate to insure quality.

An efficient and comprehensive program of Quality Assurance and Quality Control employs a number of specialized techniques adapted for each stage of Design/Construction from Concept through Certificate of Occupancy.

In this "stem to stern" approach, the Owner is assured of getting a structure whose Design incorporates Construction realities and whose Construction methods comport with those assumed during Design.

INTRODUCTION:

The processes by which quality is achieved on modern construction involves two categories of effort. Quality Assurance refers to the checking of the end product to verify its compliance with a set of established quality criteria. Quality Control refers to those activities whereby the efforts and inputs to the Design/Construction process are established or modified to produce compliance with the criteria.

The traditional approaches to quality in construction involve designing to meet pertinent codes, carefully wording specifications to close "loopholes", and a program of careful inspection, often undertaken by an independent agency engaged solely for this purpose.

The traditional results have been a steady increase in both the frequency and severity of emergent construction problems, unanticipated structural behavior, and partial or total collapse. Clearly, the "system" is not working in all cases.

Coupled with the "traditional results", and perhaps responsible for them to some degree, are the "traditional attitudes" which characterize the practice of Design and Construction in the U.S. and elsewhere. Most prevalent is the idea that "design" and "construction" are mutually

*Principal, James M. Hinckley Associates
Warwick, New York

exclusive, each with its own practitioners and standards. The engagement of the practitioners of one in the practice of the other is considered to be at least a breach of propriety and sometimes a legal liability.

Owners, who indirectly set many of the standards by which design and construction is practiced, believe that embarking on a major project is something of a crapshoot. Problems are perceived to occur with some statistical frequency and are largely unavoidable beyond the traditional care exercised in design and inspection. No real conviction exists that problems CAN be avoided.

All involved seem to feel that responsibility is an assignable commodity which, once assigned, will have no impact on those "not responsible."

If the results of efficient and comprehensive QA/QC programs can be widely documented, these attitudes may be changed and the exponential trend in problems and failures reversed.

REBUTTAL OF TRADITIONAL ATTITUDES:

The majority of construction problems occur when either:

* the Design fails to account for the realities of Construction

or

* the Construction practices do not reflect the assumptions and dictates of the Design

For designers who have not built and builders who have not designed, the process of producing and building a modern structure of spaceage materials without spaceage problems is a formidable task. The more detailed the knowledge of one practitioner in the art of the other, the better the resulting structure in terms of compatibility, durability, and service.

Most construction risks, ONCE PERCEIVED are measurable. The majority of these are avoidable. The balance of the risks have almost certainly erupted somewhere to someone and are discoverable.

Interpretations of responsibility are currently undergoing considerable change. Immediate responsibility for normal construction activities is usually defined to some extent. Contingent responsibility, that which involves reaction to a hypothetical series of events, should be assigned when perceived. Ultimate responsibility, which is assigned only after it's too late, belongs to EVERYBODY involved.

EFFICIENTLY ACHIEVING QUALITY CONSTRUCTION:

Quality construction is achieved when potential problems and risks are perceived ahead of time and steps taken to avoid them. This perception of risk is the key to an effective QA/QC program.

The author has had good results with engineering students using a simple exercise in the classroom. The class is broken into teams. Each team designs a structure in steel, concrete, or wood. At mid semester, the projects are exchanged between teams and the class is told that each design has failed. It is their responsibility to investigate the structure received during the exchange to determine the cause of the failure. The object is to instill a practice of evaluating a design by asking "What if?" to a series of hypothetical problems. The refinements are suprising for inexperienced designers, and gratifying for all involved.

In the "real world", guarding against ALL such potential problems is clearly impractical and uneconomical. The following is offered as a more realistic general approach:

1. Discover potential problems by asking "What if?" and evaluating the structural and economic impact.
2. Carefully and realistically assess the probability of occurance of circumstances creating the problem.
3. Devise the best means for repair should the problem occur, and estimate its cost.
4. Estimate the costs of preventing the problem from ocurring at all.

The Risk is then calculated:

Risk = (Economic Impact + Repair Cost) X Probability of Occurance

If the Risk Cost exceeds the cost of preventing the problem, prevent it. If the Risk Cost is equal to or slightly less than the cost of prevention, monitor the condition carefully, and act if and when the Probability of Occurance shifts the Risk Cost above the cost of prevention. If the Risk Cost is much less than the cost of avoidance, then note the potential problem for later documentation.

Several techniques are available which permit an on-going QA/QC program for each stage of the Design/Construction process. These are examined below.

QA/QC DURING CONCEPT DEVELOPMENT:

At the concept stage, any and all information obtainable about the site or sites should be collected. The methods used should be as thorough as possible, since even boring data may not reveal details which would later prove disasterous, such as caverns or fault patterns. Of particular interest is the cataloging of existing obstructions, both natural (groundwater, ledge, etc.) and man-made (utilities, buried structures, adjacent buildings, etc.). The site, if properly assessed, will pose several advantages and disadvantages. A thorough evaluation of these will permit not only the prevention of problems, but opportunities for unique solutions as well.

The structure itself may be generally evaluated for complexity and redundancy. In order to speed the structural analysis, a highly complex structure may require simplifying to such a degree that dangerous inaccuracies may result. The provision of redundancy in the arrangement of structural members, on the other hand, provides for stress redistribution and prevents a problem in one portion of the structure from spreading to other areas.

Consideration should be given to the materials to be

used. All materials have certain advantages. Inherently, therefore, all materials must have certain disadvantages. If the latter are not understood as thoroughly as the former, long-term durability and serviceability may be affected. Certain steels, for example, have extremely high strengths. The "trade off" is in ductility and resistance to even mildly corrosive atmospheres. Special compensating measures are required for their safe use.

The author has found that the keeping of a project "Design Journal" which records the design deliberations and decisions at each stage of design and construction is extremely useful. The journal records, for each of the stages of the project:

1. Potential Problems discovered.
2. The "Risk" analysis discussed above.
3. The impact of the concept or design on later stages of design, analysis, and construction.

As the journal develops, it produces a compendium of information which is ultimately formalized as a "Manual" at the completion of the project.

DESIGN DEVELOPMENT:

Design Development should address the problems postulated in the Concept phase. In addition, other areas become more defined and may be, in turn, subjected to the same "What if?" consideration.

The materials to be used become more clearly specified. For each of the newer materials or those being employed in non-traditional ways, a list of limitations should be compiled. Means of overcoming each of these limitations is developed and the result is a series of design guidelines to assure proper application of newer products and components. This list is also useful in developing the project specifications.

Related to this issue is the question of the compatibility of adjacent materials. Catalytic reactions between dissimilar metals are well-known, and are usually avoided. With other materials, adverse reactions are not as well recognized, and should be documented in advance to guide the final design.

A "buildability analysis" was unheard of not so long ago. With structures becoming increasingly complex, this may represent an essential step to avoid costly delays during erection. With several designers working on various areas of a structure, incompatibilities between areas is not unheard of. If the steel connection in Phase 2 of construction cannot be made because of an adjacent concrete slab in Phase 1, the costs of redesign and delays to the project will dwarf the costs of avoiding this problem during the design development.

An often overlooked consideration is that of maintenance and periodic inspection of the structure by the owner. The investigations of many failures have documented the need for access to all areas for inspection and routine maintenance.

If this is given consideration as the design is developed, costly retrofitting, or worse, can be avoided.

When the design development is complete, the construction sequencing should be planned. This will assure that the components of the building which will be only partially supported during erection will be stable and safe until erection is complete. This will also help to spot problems of buildability, since at least ONE way of assembling the structure will be known in advance of contract letting.

STRUCTURAL ANALYSIS AND DESIGN:

The detailed analysis of any structure involves the use of assumptions to simplify and speed the process. Concepts which have no physical meaning, such as the frictionless pin, are routinely employed and, in most cases, conservatively so. The analysis of large redundant frames, however, may be rendered sufficiently in error by such methods that serious under-designs of members or connections may result. The accuracy of all simplifying assumptions in the specific design application must be known, therefore, to avoid a misleading analytical result.

It is recommended that the use of computer methods of analysis be similarly checked for accuracy on a particular structure, unless experience with this program on this type of structure exists. The check can consist of running a simple subassembly of the structure, and comparing the results of hand analysis.

Following the analysis, the materials to be employed should be given a final check to assure their suitability for the intended use. The effect of fluctuating stress levels, vibrations, atmospheric exposure, and secondary stress patterns are typical of those included in this final verification.

Finally, the structure in its various stages of partial assembly, as revealed by the construction sequence assumed during design development, should be analyzed to verify stability during erection and to use as a basis for comparison with actual measurements taken during the construction process.

SHOP DRAWING GUIDELINES AND REVIEW:

The review of shop drawings has taken on national prominence both within the engineering profession and outside of it. The importance has been stressed but, to date, no system of guidelines has been offered.

To facilitate the review of shop drawings, it would be well to adopt a unified set of guidelines governing their preparation and the information shown. As a first step, consider the classification of building components into three categories. Catagory One would contain those elements which EXERT a load, but do not support them, such as roof gravel, low profile mechanical equipment, etc. Category Two is comprised of elements which EXERT loads, but support only their own weight and the forces of nature, such as curtain wall panels, glazing, etc. Category Three includes those elements which both EXERT loads and SUPPORT the loads from other elements, such as concrete, structural steel, etc.

If drawings and specifications for Category Two and Three components are sealed by an engineer, one concern of the designer has been addressed.

It makes review much easier if the drawings also clearly indicate the points of attachment to adjacent elements fabricated "by others" and show the loads exerted at those points of attachment.

In addition, the materials submitted by each vendor should be reviewed by the vendor to which his elements connect, and bear the approval stamp of the latter that the supporting elements have been designed for those loads.

The designer, of recent ruling, is responsible for the "working" of the whole, but the exchange of drawings between vendors helps to assure that no major item is overlooked.

FABRICATION AND SUBASSEMBLY:

For most fabricated components in construction, there are several methods of manufacture. Unless the method is specified, which is rare, the component produced may or may not achieve the results desired. It is not enough to simply specify the technical parameters of the finished product, since some methods of manufacture may result in flaws not covered in the specs. For this and other reasons, it may be desirable to research the methods of manufacture and thoroughly test specimen components from a full production run prior to approving the vendor's product. Flaws in structural castings, for example, may be related to shell mold design and be overlooked by some of the common methods of inspection. Production specimens, as contrasted with samples specifically cast as samples, are more likely to be representative of the delivered components.

Errors may occur in fabrication, subassembly, or in inspection. These errors are of two general types. Errors of Commission are those which result when some operation was done improperly. Errors of Omission, on the other hand, are the result of an unintentioned oversight. An error of commission is to clamp the end of a cable with the clamp on backwards. An error of omission is to check ten of the fifteen clamps in an assembly and, finding them properly installed, certify the assembly as approved.

The inspection program that is established must, therefore, guard against both types of errors. The program must specify:

1. The parameters to be tested.
2. The frequency of testing (i.e. the number of units to be tested per 100 units produced).
3. The means of testing.
4. The measurement readings considered acceptable.
5. The means of reporting the results of the tests.

The parameters should be based on the specifications and whatever deficiencies were discovered in the process of verifiying the method of fabrication. The frequency of testing should begin very high, with the first units from production subjected to as high as 100% testing. As experience is gained and acceptable test results are obtained, the frequency may be reduced. If test results deteriorate, provision must be made for increasing the frequency of testing. The means of testing should be as thorough as possible for the first units of the production run, with a reservation for reducing the "severity" of the tests as acceptable results are received. The permissable tolerances should be realistic, but based on a possibility that, depending upon the testing method, conditions worse than the allowable tolerances may slip through. Finally, the reporting system used should enable anyone following up on the inspection reports to trace the production and testing path of ANY particular component. This will require identification by unit number, pattern, heat number, etc. and the inspection report for each tested component must show these identification numbers. Such a system has several times kept the American public from being decimated by canned soups and non-prescription remedies. Its applicability here is clear.

The inspectors must be equipped with a means of marking and permanently identifying components which have been accepted and those rejected. The markings should be clearly discernable and the absence of either marking should indicate that the component has not been inspected.

ERECTION:

Methods of erection should be COMPATIBLE, if not the same, as those anticipated during the Concept and Design Development phases. If the assembly is unique, the crews relatively inexperienced with the assembly procedures, or the site conditions considered adverse, a mockup may be prudent. The procedures may then be evaluated for efficiency and the resulting assembly tested for compliance with the specifications and the dictates of good workmanship.

The assembly process should be thoroughly planned and sequenced, to assure that the partially-erected structure will withstand all live and dead loads imposed during construction. Many collapses are the result of imposing "construction" loads greatly in excess of design live loads, or erecting a partial assembly that subsequently buckled due to inherent instability.

Recommendations above for the development of a fabrication inspection program should also apply to the inspection of field erections. Here, tolerances may be even more critical than in the shop, where controls over manufacture and subassembly are more precise. Approved and rejected portions of the erected structure should also be clearly marked, so that the work of following trades will not be applied to a rejected area.

As the structure is erected, it may be important to actually monitor the deflections of key locations and compare them to the analyses of the partially-erected structure run during the Structural Analysis phase. If something is amiss, it should show up as an excess deflection. This process enables the engineer to feel the "pulse" of the structure throughout construction.

Change orders during construction are all too often evaluated solely upon consideration of the immediately-affected construction. Any change to the structural components of a project has the potential of affecting other, often remote, portions of the structure. Thorough evaluation of these potentials should preceed the approval of structural changes, however slight they may appear. The decision to approve the use of understrength concrete, for example, should be made after reviewing not only the design live loads, but the construction loads as well.

FINAL INSPECTION AND MONITORING:

Before scheduling the Final Inspection, a review should be made of the Journal kept by the QA/QC personnel, and any remaining issues resolved. The contents of this Journal, suitably edited, can then be made the basis for preparing an Owner's Operation, Inspection, and Maintenance Manual for the structure. This will detail the design loads, locations and frequencies of routine inspection, and a program of recommended maintenance to assure structural reliability. Such a practice is being adopted currently by some state transportation officials and should be expanded to include engineered structures of every type.

Final Inspection should include verification that all "open items" have been met, and that the building is ready for beneficial occupancy.

On large or complex structures, a program of structural monitoring may be desirable, to verify conformance to the predictions of behavior made during Structural Analysis. Any significant variations from this standard should be investigated thoroughly and verification made of the soundness of the constructed work. This program should extend through at least one year, so that seasonal variations in natural loading conditions, "creep" effects, and other time-related phenomena are studied.

CONCLUSION:

Quality Assurance/Quality Control on modern construction is today a critical function, as projects become more complex, lighter, more expensive, and harder to track. A creative staff of engineers, given a certain degree of freedom to think, can anticipate and avoid many of the pitfalls currently plaguing those projects which are without such an effort. Far from being a "crapshoot", the tools of the construction industry which are now available offer a greater opportunity than ever before to reduce the frequency, severity, and economic impact of construction problems and failures.

SUBJECT INDEX

Page number refers to first page of paper.

AUTHOR INDEX

Page number refers to first page of paper.